SpringerBriefs in Applied Sciences and Technology

Nonlinear Circuits

Series editors

Luigi Fortuna, Catania, Italy
Guanrong Chen, Kowloon, Hong Kong SAR, P.R. China

SpringerBriefs in Nonlinear Circuits promotes and expedites the dissemination of substantive new research results, state-of-the-art subject reviews and tutorial overviews in nonlinear circuits theory, design, and implementation with particular emphasis on innovative applications and devices. The subject focus is on nonlinear technology and nonlinear electronics engineering. These concise summaries of 50–125 pages will include cutting-edge research, analytical methods, advanced modelling techniques and practical applications. Coverage will extend to all theoretical and applied aspects of the field, including traditional nonlinear electronic circuit dynamics from modelling and design to their implementation. Topics include but are not limited to:

- nonlinear electronic circuits dynamics;
- Oscillators;
- cellular nonlinear networks;
- arrays of nonlinear circuits;
- chaotic circuits;
- system bifurcation;
- chaos control;
- active use of chaos;
- nonlinear electronic devices;
- memristors;
- circuit for nonlinear signal processing;
- wave generation and shaping;
- nonlinear actuators;
- nonlinear sensors;
- power electronic circuits;
- nonlinear circuits in motion control;
- nonlinear active vibrations;
- educational experiences in nonlinear circuits;
- nonlinear materials for nonlinear circuits; and
- nonlinear electronic instrumentation.

Contributions to the series can be made by submitting a proposal to the responsible Springer contact, Oliver Jackson (oliver.jackson@springer.com) or one of the Academic Series Editors, Professor Luigi Fortuna (luigi.fortuna@dieei.unict.it) and Professor Guanrong Chen (eegchen@cityu.edu.hk).

More information about this series at http://www.springer.com/series/15574

Qianxue Wang · Simin Yu
Christophe Guyeux

Design of Digital Chaotic Systems Updated by Random Iterations

 Springer

Qianxue Wang
School of Automation
Guangdong University of Technology
Guangzhou, Guangdong
China

Christophe Guyeux
Femto-st institute, UMR 6174 CNRS
University of Bourgogne Franche-Comté
Besançon
France

Simin Yu
School of Automation
Guangdong University of Technology
Guangzhou, Guangdong
China

ISSN 2191-530X   ISSN 2191-5318   (electronic)
SpringerBriefs in Applied Sciences and Technology
ISSN 2520-1433   ISSN 2520-1441   (electronic)
SpringerBriefs in Nonlinear Circuits
ISBN 978-3-319-73548-1   ISBN 978-3-319-73549-8   (eBook)
https://doi.org/10.1007/978-3-319-73549-8

Library of Congress Control Number: 2017964002

Mathematics Subject Classification (2010): 34C28

Printed on acid-free paper

This Springer imprint is published by the registered company Springer International Publishing AG
part of Springer Nature
The registered company address is: Gewerbestrasse 11, 6330 Cham, Switzerland

Preface

This monograph is an extended version of the postdoctoral report guided by Professor Simin Yu [1], completed in the postdoctoral training program of control science and engineering at the Guangdong University of Technology in October 2015. Chapters 2, 3, and 5 are reproduced with permission from papers [2, 3, 4] with additional Sects. 2.2.3, 2.2.4, 2.3, 3.3, 5.1.1, and 5.3.3.

The underlying theme of the monograph is to build digital chaotic systems updated by random iterations on the domain of finite precision from low- to high-dimensional settings. Traditionally, chaotic systems are built on the domain of infinite precision in mathematics. However, quantization is inevitable for digital devices, which causes dynamical degradation. To cope with this problem, we study the general problem of constructing digital chaotic systems updated by random iterations in digital devices with finite precision from low-dimensional to high-dimensional settings, and we establish a general framework of composing them. We demonstrate that the associated state network of our digital chaotic systems are strongly connected, and then further prove that such systems updated by random iterations satisfy Devaney's definition of chaos in the domain of finite precision. In addition, Lyapunov exponents are also given, and implementations are then provided to show the potential application of our digital chaotic systems updated by random iterations in the digital world. The basic advantage and practical benefits of the proposed approach are that chaos satisfying Devaney's definition can be generated in the real digital devices by means of a chaos generation strategy controlled by random sequences.

The core of the monograph is the chaos generation strategy controlled by random sequences of our digital chaotic systems: through a random control sequence, some bits are randomly updated by the iterative equation when the remaining bits keep their original value in each updating iteration. This is also called the iterative update mechanism controlled by random sequences. Our digital chaotic systems updated by random iterations operate in a totally different way from traditional real domain chaotic systems, whose bits are all updated by an iterative equation at each iteration.

In 2010, Bahi et al. proposed a 1D integer domain chaotic system (IDCS) designated as a chaotic iterations (CI) system [5]. Since then, there has been a lot of attention to digital chaotic systems in the domain of finite precision and for their applications. CI systems have been refined and applied to various important new examples in [6]. Writing a new monograph from scratch today, we take as a starting point the notion of digital chaotic systems, and proceed to their applications. References are given to relevant papers so that the interested reader may follow the details. We hope that the monograph will provide a useful introduction to what promises to be an exciting and fruitful research area.

We wish to record our warmest thanks to Professors Jacques M. Bahi, Guanrong Chen, Jinhu Lü, and Chengqing Li for their numerous suggestions and discussions during the development of this work and for reading everything critically. We would also like to thank Xiaole Fang for helpful discussions, and our friends and colleagues at the FEMTO-ST Institute in France and at the Guangdong University of Technology in China for providing a pleasant and stimulating research environment.

Guangzhou, China   Qianxue Wang
Guangzhou, China   Simin Yu
Besançon, France   Christophe Guyeux
May 2017

References

1. Q. Wang, *Study on Modeling, Analyze and Application of Integer Domain and Digital Domain Chaotic System*. Postdoctoral report (Guangdong University of Technology, China, 2015), p. 85
2. Q. Wang, S. Yu, C. Guyeux, J. Bahi, X. Fang, Theoretical design and circuit implementation of integer domain chaotic systems. Int. J. Bifurc. Chaos **24**(10), Art. No. 1450128 (2014)
3. Q. Wang, S. Yu, C. Guyeux, J. Bahi, X. Fang, Study on a new chaotic bitwise dynamical system and its FPGA implementation. Chin. Phys. B **24**(6), Art. No. 60503 (2015)
4. Q. Wang, S. Yu, C. Li, J. Lü, X. Fang, C. Guyeux, J. Bahi, Theoretical design and FPGA-based implementation of higher-dimensional digital chaotic systems. IEEE Trans. Circuits Syst. I **63**(3), 401–412 (2016)
5. C. Guyeux, J. Bahi, Hash functions using chaotic iterations. J. Algorithm Comput. Technol. **4**(2), 167–182 (2010)
6. J. Bahi, C. Guyeux, *Discrete Dynamical Systems and Chaotic Machines: Theory and Applications* (Chapman and Hall/CRC, 2013), p. 212

Acknowledgements

This work was supported by the National Key Research and Development Program of China (No. 2016YFB0800401), the National Natural Science Foundation of China (No. 61532020, 61671161, 61172023), and the Science and Technology Planning Project of Guangzhou (No. 201510010136).

Contents

1 An Introduction to Digital Chaotic Systems Updated by Random Iterations 1
  1.1 General Presentation 1
  1.2 Mathematical Definitions of Chaos 3
    1.2.1 Approaches Similar to Devaney 3
    1.2.2 Li–Yorke Approach 5
    1.2.3 Topological Entropy Approach 5
    1.2.4 Lyapunov Exponent 6
  1.3 TestU01 6
  1.4 Plan of This Book 7
  References 8

2 Integer Domain Chaotic Systems (IDCS) 11
  2.1 Description of IDCS 11
    2.1.1 Real Domain Chaotic Systems (RDCS) 11
    2.1.2 IDCS 12
  2.2 Proof of Chaos for IDCS 16
    2.2.1 Dense Periodic Points 16
    2.2.2 Transitive Property 17
    2.2.3 Further Investigations of the Chaotic Behavior
          of IDCS 19
    2.2.4 Relationship Between Iterative Input and Output 20
  2.3 Network Analysis of the State Space of IDCS 22
    2.3.1 The Corresponding State Transition Diagram
          and Its Connectivity Analysis for IDCS with $N = 3$ ... 22
    2.3.2 The Corresponding State Transition Diagram
          and Its Connectivity Analysis for IDCS with $N = 4$ ... 24
  2.4 Circuit Implementation of IDCS 26
  References 33

3   Chaotic Bitwise Dynamical Systems (CBDS)   35
  3.1   Improvements of Chaotic Bitwise Dynamical Systems
      (CBDS)   35
  3.2   Proof of Chaos for CBDS   38
     3.2.1   Dense Periodic Points   38
     3.2.2   Transitive Property   39
  3.3   Uniformity   40
  3.4   TestU01 Statistical Test Results   42
  3.5   FPGA-Based Realization of CBDS   43
  References   45

4   One-Dimensional Digital Chaotic Systems (ODDCS)   47
  4.1   The Structure of One-Dimensional Digital Chaotic Systems   47
     4.1.1   The Conventional Iterative Update Mechanism   47
     4.1.2   The Iterative Update Mechanism Controlled
        by Random Sequences   48
  4.2   The Connection Between a Chaotic System and Its Strongly
      Connected Network   50
     4.2.1   Transitive Property of ODDCS   51
     4.2.2   Dense Periodic Points of ODDCS   52
     4.2.3   Chaotic System and Its Strongly Connected Network   53
  4.3   Lyapunov Exponents of a Class of ODDCS   53
     4.3.1   General Expression of Equivalent Decimal for G_F   53
     4.3.2   Mathematical Expression for $\frac{\partial G(y)}{\partial y}$ 54
     4.3.3   Estimating the Lyapunov Exponents   55
  Reference   57

5   Higher-Dimensional Digital Chaotic Systems (HDDCS)   59
  5.1   Design of HDDCS   59
     5.1.1   Higher-Dimensional Integer Domain Chaotic Systems
        (HDDCS)   59
     5.1.2   Description of HDDCS   61
     5.1.3   Comparative Study of RDCS, IDCS, CBDS,
        and HDDCS   65
     5.1.4   Network Analysis of the State Space of HDDCS   68
  5.2   Chaotic Performance of HDDCS   72
     5.2.1   Dense Periodic Points of HDDCS   72
     5.2.2   Transitive Property of HDDCS   75
  5.3   Lyapunov Exponents of a Class of HDDCS   78
     5.3.1   General Expression of Equivalent Decimal for G_F   78
     5.3.2   Mathematical Expression for $\frac{\partial g_i(y_1,y_2,...,y_m)}{\partial y_j}$ 80
     5.3.3   Estimating the Lyapunov Exponents   81

5.4   FPGA-Based Real-Time Application of 3D-DCS   84
        5.4.1   Design of 3D-DCS in FPGA   84
        5.4.2   Design of the FPGA-Based Hardware System
                for Image Encryption and Decryption   85
        5.4.3   FPGA-Based Implementation Result for Image
                Encryption and Decryption   88
References   88

6   **Investigating the Statistical Improvements of Various Chaotic
    Iterations-Based PRNGs**   89
    6.1   Various Algorithms for Pseudorandom Number Generation   89
            6.1.1   Qualitative Relations Between Topological Properties
                    and Statistical Tests   89
            6.1.2   CIPRNGs: Chaotic Iteration-Based PRNG Algorithms   91
    6.2   On the Statistical Improvements of CIPRNG Posttreatments   95
            6.2.1   First Investigations   95
            6.2.2   Variations on the XOR CIPRNG   97
            6.2.3   "LUT" CIPRNG (XORshift, XORshift) Version 3   98
            6.2.4   The Version 4 Category of CIPRNGs   98
            6.2.5   Randomness Quality of CIPRNGs   99
    6.3   Practical Security Evaluation   100

7   **Conclusions**   103

Appendix A: Some Well-Known Generators   105

Abbreviations

CBDS   Chaotic bitwise dynamical system
HDDCS   Higher-dimensional digital chaotic system
IDCS   Integer domain chaotic system
ODDCS   One-dimensional digital chaotic system
PRNG   Pseudorandom number generator
RDCS   Real domain chaotic system
RNG   Random number generator
TRNG   True random number generator

Chapter 1
An Introduction to Digital Chaotic Systems Updated by Random Iterations

1.1 General Presentation

Usually, in the literature, only the real domain is investigated when designing and studying chaotic systems. Such research works are grouped under the name of real domain chaotic systems (RDCSs); they mainly encompass two distinct approaches: the continuous-time systems and discrete-time chaotic ones. The former are constituted by differential state equations, whereas the latter are based on iterative equations. Well-known Chen, Chua, and Lorenz systems [1–4] belong to the first category, and logistic or Henon maps, or Chen-Lai algorithms [5–8] are well-known examples of the second category.

It is unnecessary to exhibit a complex structure for a chaotic system to have, mathematically speaking, a complex dynamics: expansivity, sensitivity to initial conditions, topological mixing, decaying autocorrelation function, and so on [5, 9, 10]. And, there is no denying that these properties may be related to some desired aspects of information security. For instance, key sensitivity for the avalanche effect, in symmetric ciphers that are iterative systems, seems to be related to the sensitivity to the initial conditions for chaotic systems [11–14]. Such remarks explain why chaos-based encryption schemes have recently been considered as interesting ways to investigate, as it may allow a security reinforcement, at least against auxiliary channel hardware attacks [15–23].

There is, however, a real problem with such an approach, related to the implementation of a chaotic phenomenon, no matter the considered device: the associated chaotic system is discretized both spatially and temporally. In other words, the system becomes both a discrete-valued and discrete-time "pseudochaotic" system, on a finite spatial lattice and for discrete times [24–26]. In this book, by digital chaos, we mean the exact result of these discretized iterations on digital computers.

In digital finite state machines, finite word lengths imply finite precision. This may lead to important dynamic degradation: nonideal distribution and correlation, short cycle length, low linear complexity, and so on [27, 28]. In addition, important

Q. Wang et al., *Design of Digital Chaotic Systems Updated by Random Iterations*, SpringerBriefs in Nonlinear Circuits, https://doi.org/10.1007/978-3-319-73549-8_1

security issues may result in a degradation of various extents of an implementation of a digital chaotic system in finite-precision devices; see, for example, [29–36]. Furthermore, division, multiplication, or any other linear or nonlinear operation over the real numbers may raise quantization, rounding, or overflow errors. This will lead to large differences between actual orbits and theoretical chaotic ones [37, 38]. Thus, in order to face such degradation, various approaches have been proposed, such as disturbing the control parameter [39–41], using a larger finite precision [42], perturbing chaotic states [40, 43, 44], or operating a cascading of two or more chaotic maps [45–47]. However, Hu and Deng have recently demonstrated that the phase space of a chaotic system should not be implemented in infinite precision [33], whereas mathematical chaos definitions such as the reputed Devaney's one are compatible with finite precision domains.

Even though numerous research works have recently been published, targeting a disclosure of the degradation characteristic for some simple digital domain chaotic maps (including the piecewise linear chaotic [27] or the logistic maps [48]), there still remains a lack of systematic theoretical investigations that rigorously analyze the degradation caused by finite precision effects. This is why Guyeux et al. proposed a one-dimensional digital chaotic system updated by random iterations designated as chaotic iterations (CIs) in 2010 [49]. The chaotic iterations-based system works as follows. It takes one or two random stream(s) as input. It then operates their ad hoc mix, where the involved mixture function is only constituted by binary operations. The resulting discrete dynamical system has been theoretically proven to satisfy the mathematical definition of chaos, as proposed by Devaney [50]. Obviously, the proposed system only manipulates finite sets of integers, therefore the issue related to finite precision does not occur. Indeed, as there is no need to transform real numbers to binary sequences, CIs-based systems can be considered one of the most effective solutions to resolve the aforementioned degradation problem in digital chaotic systems.

In more detail, the original work consisted of a chaotic combination of two random inputs leading to a first CIs-based system named PRIM CI, (chaotic iteration; see [51]). After having studied it, we showed that the output presents better statistical properties than each input taken alone. We then proposed in [52] another category of chaotic iterations-based systems called MARK CI. In this latter, a mark sequence was used to avoid wasteful duplication of values, thus leading to a speed improvement. Finally, the lookup-table (LUT) CIs-based pseudorandom number generator was proposed and studied in [53]. In this last version of the chaotic combination of two inputted entropic streams, we solved all the flaws exhibited by the MARK CI version.

In this book, we study the general problem of constructing digital chaotic systems updated by random iterations from low-dimensional to high-dimensional settings, and establish a general framework of composing them. Our digital chaotic systems updated by random iterations are classified into two types: ODDCS(one-dimensional digital chaotic systems, including integer domain chaotic systems, IDCS, and chaotic bitwise dynamical systems, CBDS, in this category) and HDDCS (higher-dimensional digital chaotic systems). We demonstrate that the associated state networks of our systems are strongly connected, and then further prove that they satisfy the Devaney definition of chaos on the domain of finite precision. In addition, Lyapunov exponents for such systems are also given, analog-digital hybrid circuit and FPGA-based implementations are then provided to show the potential applications of digital chaotic systems updated by random iterations in the digital world. The basic advantage and practical benefits of the proposed approach is that the chaos satisfying Devaney's definition can be generated in the real digital devices by means of the chaos generation strategy controlled by random sequences.

As we intend to use chaos of pseudorandom number generation, we need first to define what chaos is, and how to check the randomness of generated sequences. This is the objective of the two following sections, and this introductory chapter ends with the plan of this book.

1.2 Mathematical Definitions of Chaos

For the sake of completeness, let us first recall various key definitions and properties in the field of the mathematical theory of chaos.

In this section, to prevent any conflicts and to avoid unreadable writings, we have considered the following notations, usually used in discrete mathematics:

- The nth term of the sequence s is denoted by s^n.
- The ith component of vector v is v_i.
- The kth composition of function f is denoted by f^k. Thus $f^k = f \circ f \circ \ldots \circ f$, k times.
- The derivative of f is f'.

$\mathbb{B}$ stands for the set $\{0; 1\}$ with its usual algebraic structure (Boolean addition, multiplication, and negation), whereas $\mathbb{N}$ and $\mathbb{R}$ are the usual notations of the respective sets, natural numbers, and real numbers. $\mathscr{X}^{\mathscr{Y}}$ is the set of applications from $\mathscr{Y}$ to $\mathscr{X}$, and thus $\mathscr{X}^{\mathbb{N}}$ means the set of sequences belonging in $\mathscr{X}$. We use the notation $\lfloor x \rfloor$ for the integral part of a real x, that is, the greatest integer lower than x. Finally, $[\![a; b]\!] = \{a, a+1, \ldots, b\}$ is the set of integers between a and b.

Various explanations of a chaotic behavior for a discrete dynamical system are presented in the following.

1.2.1 *Approaches Similar to Devaney*

In these approaches, certain ingredients are required for unpredictability. First, the system must be intrinsically complicated and undecomposable: it cannot be simplified into two subsystems that do not interact, making any divide and conquer strategy applied to the system inefficient. In particular, many orbits must visit the whole space. Second, an element of regularity is added to counteract the effects of the first ingredient, leading to the fact that closed points can behave in a completely different manner, and this behavior cannot be predicted. Finally, sensibility of the system is demanded as a third ingredient, enabling close points finally to become distant during iterations of the system. This last requirement is, indeed, often implied by the two first ingredients. Having this understanding of an unpredictable dynamical system, Devaney has formalized the following definition of chaos.

Definition 1.1 (*Devaney's definition of chaos* [50]) A discrete dynamical system $x^0 \in \mathcal{X}, x^{n+1} = f(x^n)$ on a metric space $(\mathcal{X}, d)$ is chaotic according to Devaney if:

1. *Transitivity:* For each couple of open sets $A, B \subset \mathcal{X}$, there exists $\exists k \in \mathbb{N}$ such that $f^k(A) \cap B \neq \varnothing$.
2. *Regularity:* Periodic points are dense in $(\mathcal{X}, d)$.
3. *Sensibility to the initial conditions:* There exists $\varepsilon > 0$ such that

$$\forall x \in \mathcal{X}, \forall \delta > 0, \exists y \in \mathcal{X}, \exists n \in \mathbb{N}, d(x, y) < \delta \text{ and } d(f^n(x), f^n(y)) \geqslant \varepsilon.$$

Let us recall that

Theorem 1.1 (Banks [54]) *If a dynamical system is transitive and has dense periodic points on a metrical space, then it has sensitive dependence on initial conditions.*

Thus, to prove that we are in the framework of Devaney's topological chaos, we simply have to check the two conditions of regularity and transitivity. The system can be intrinsically complicated for various other explanations of this wish, that are not equivalent to one another, such as:

- *Undecomposable*: It is not the union of two nonempty closed subsets that are positively invariant ($f(A) \subset A$).
- *Total transitivity*: $\forall n \geqslant 1$, the function composition f^n is transitive.
- *Strong transitivity*: $\forall x, y \in \mathcal{X}, \forall r > 0$, there $\exists z \in B(x, r)$, and there $\exists n \in \mathbb{N}$, $f^n(z) = y$.
- *Topological mixing*: For all pairs of disjoint open nonempty sets U and V, there exists $n_0 \in \mathbb{N}$ such that $\forall n \geqslant n_0, f^n(U) \cap V \neq \varnothing$.

Concerning the ingredient of sensibility, it can be reformulated as follows.

- $(\mathcal{X}, f)$ is *unstable* if all its points are unstable: $\forall x \in \mathcal{X}$, there $\exists \varepsilon > 0, \forall \delta > 0$, there $\exists y \in \mathcal{X}$, and there $\exists n \in \mathbb{N}, d(x, y) < \delta$, and $d(f^n(x), f^n(y)) \geqslant \varepsilon$.
- $(\mathcal{X}, f)$ is *expansive* if there $\exists \varepsilon > 0, \forall x \neq y$, and there $\exists n \in \mathbb{N}, d(f^n(x), f^n(y)) \geqslant \varepsilon$

These varieties of definitions lead to various notions of chaos. For instance, a dynamical system is chaotic according to Wiggins if it is transitive and sensible to the initial conditions. It is said to be chaotic according to Knudsen if it has a dense orbit while being sensible. Finally, we speak about expansive chaos when the properties of transitivity, regularity, and expansiveness are satisfied.

1.2.2 Li–Yorke Approach

The approach for chaos presented in the previous section, considering that a chaotic system is intrinsically complicated (undecomposable), with possibly an element of regularity and/or sensibility, and has been completed by other perceptions of chaos. Indeed, as "randomness" or "infinity," a single universal definition of chaos cannot be found. The kind of behaviors that are attempted to be described are too complicated to comprise only one definition. Instead, in the last decades a large panel of mathematical descriptions has been proposed, all being theoretically justified. Each of these definitions illustrates some particular aspects of a chaotic behavior.

The first of these parallel approaches can be found in the pioneer work of Li and Yorke [55]. In their well-known article entitled "Period Three Implies Chaos," they rediscovered a weaker formulation of Sarkovskii's theorem, meaning that when a discrete dynamical system $(f, [0, 1])$, with f continuous, has a 3-cycle, then it also has an $n-$cycle, $\forall n \leqslant 2$. The community has not adopted this definition of chaos, as several degenerated systems satisfy this property. However, in their article [55], Li and Yorke have studied another interesting property, which has led to a notion of chaos "according to Li and Yorke" recalled below.

Definition 1.2 Let $(\mathscr{X}, d)$ be a metric space and $f : \mathscr{X} \longrightarrow \mathscr{X}$ a continuous map. $(x, y) \in \mathscr{X}^2$ is a scrambled couple of points if $\lim \inf_{n \to \infty} d(f^n(x), f^n(y)) = 0$ and $\lim \sup_{n \to \infty} d(f^n(x), f^n(y)) > 0$. In other words, the two orbits oscillate.

A scrambled set is a set in which any couple of points are a scrambled couple, whereas a Li–Yorke chaotic system is a system possessing an uncountable scrambled set.

1.2.3 Topological Entropy Approach

Let $f : \mathscr{X} \longrightarrow \mathscr{X}$ be a continuous map on a compact metric space $(\mathscr{X}, d)$. For each natural number n, a new metric d_n is defined on $\mathscr{X}$ by

$$d_n(x, y) = \max\{d(f^i(x), f^i(y)) : 0 \leq i < n\}.$$

Given any $\varepsilon > 0$ and $n \geqslant 1$, two points of $\mathscr{X}$ are ε-close with respect to this metric if their first n iterates are ε-close. This metric allows one to distinguish in a neighborhood of an orbit the points that move away from each other during the iteration from the points that travel together. A subset E of $\mathscr{X}$ is said to be (n, ε)-separated if each pair of distinct points of E is at least ε apart in the metric d_n. Denote by $N(n, \varepsilon)$ the maximum cardinality of an (n, ε)-separated set. $N(n, \varepsilon)$ represents the number of distinguishable orbit segments of length n, assuming that we cannot distinguish points within ε of one another.

Definition 1.3 The topological entropy of the map f is defined by

$$h(f) = \lim_{\epsilon \to 0} \left(\limsup_{n \to \infty} \frac{1}{n} \log N(n, \epsilon) \right).$$

The limit defining $h(f)$ may be interpreted as the measure of the average exponential growth of the number of distinguishable orbit segments. In this sense, it measures the complexity of the topological dynamical system $(\mathscr{X}, f)$.

1.2.4 Lyapunov Exponent

The last measure of chaos regarded in this document is the Lyapunov exponent. This quantity characterizes the rate of separation of infinitesimally close trajectories. Indeed, two trajectories in phase space with initial separation δ diverge at a rate approximately equal to $\delta e^{\lambda t}$, where λ is the Lyapunov exponent, which is defined by:

Definition 1.4 Let $f : \mathbb{R} \longrightarrow \mathbb{R}$ be a differentiable function, and $x^0 \in \mathbb{R}$. The Lyapunov exponent is given by $\lambda(x^0) = \lim_{n \to +\infty} \frac{1}{n} \sum_{i=1}^{n} \ln \left| f'\left(x^{i-1}\right) \right|$.

Obviously, this exponent must be positive to have a multiplication of the initial errors by an exponentially increasing factor, and thus chaos in this interpretation.

Note that all the aforementioned notions of chaos have been studied in our previously published research work; see the references at the end of this book. However, for the sake of concision, we only focus here on the most used notions of mathematical chaos, namely the Devaney approach and Lyapunov exponent.

1.3 TestU01

There are many approaches to analyze the safety performance of chaotic systems used in information security, depending on which aspect sounds important. These approaches encompass speed, cryptographical security, or statistical profile evaluation. This latter is a prerequisite for numerous applications; it is quite quick and simple to evaluate through reputed batteries of tests. Furthermore it does not need to consider the internal structure of the chaotic system, which explains why using such batteries becomes an internationally accepted analytical tool for PRNGs.

Various statistical tests are available in the literature that test a given sequence for some level of computational indistinguishability with a truly random stream. Major test suites for random sequences are TestU01 [56], the NIST suite [57], and the DieHARD suite [58]. The DieHARD suite, which implements many classical tests, has revealed over time some drawbacks and limitations. The National Institute of Standards and Technologies (NIST) in the United States, for its part, has implemented a suite (16 tests) for testing random sequences. It is geared mainly for the testing

Table 1.1 Some batteries of tests in TestU01

Battery	Testing data	Number of statistics
Rabbit	32Gb	40
Alphabit	32Gb	17
Pseudo DieHARD	5Gb	126
FIPS_140_2	19Kb	16
Small crush	6Gb	15
Crush	973Gb	144
Big crush	10Tb	160

and certification of random sequences used in cryptographic applications. Finally, TestU01 is extremely diverse in implementing classical tests, cryptographic tests, new tests proposed in the literature, and original ones. In fact, it encompasses most of the other test suites. The TestU01 test suite has three battery levels as shown in Table 1.1.

1.4 Plan of This Book

The remainder of this book is organized as follows.

In Chap. 2, an approach for generating IDCS is presented, together with its proof of chaos according to Devaney's definition. We then focus on the design and circuit implementation of IDCS, with theoretical background and practical details presented. The IDCS circuit design consists of a uniform noise signal generator, noise voltage converter, sample and hold circuit, decoder circuit, iterative function circuit, and digital-to-analog converter: six parts altogether. The main feature of this kind of IDCS circuit is the use of a sample-hold circuit and a decoder circuit to convert the uniform noise signal into a random sequence, which plays a key role in generating IDCS signals.

In Chap. 3, a CBDS is proposed, whose dynamics of chaos is still preserved in digital computers. Its obvious speed improvement makes it a good candidate when developing new approaches in some fields regarding information security such as digital watermarking. It uses multiple random bitwise operations, instead of only one as in [59]. Proofs of chaos according to Devaney's definition are also presented in this chapter. After having introduced the theoretical framework of this study, we give a comparison based on the uniformity of the sequence and on statistical tests. We then focus on the field programmable gate array (FPGA) implementation of this CBDS, with the presentation of both theoretical background and practical details. Some specific components including the ring oscillator [60], D flip-flop, iterate function circuit, and DAC (digital-to-analog converter) have been applied to design the FPGA generator. The main feature of this kind of CBDS circuit is that it takes advantage of parallel computations in FPGA.

Then, in Chap. 4, the structure of a ODDCS in digital devices with finite precision is summarized, and a general framework for ODDCS composition is established. We demonstrate again that ODDCS satisfies Devaney's definition of chaos on the domain of finite precision if and only if the associated state network is strongly connected. Lyapunov exponents for such ODDCSs are also given.

In Chap. 5, we study the general problem of constructing higher-dimensional digital chaotic systems (HDDCSs) in digital devices with finite precision, and we establish a general framework to design HDDCSs. We demonstrate that the associated state network of HDDCSs is strongly connected, and then we prove that HDDCS satisfies Devaney's definition of chaos on the domain of finite precision. In addition, Lyapunov exponents for such a HDDCS are also given, and FPGA-based implementations are then provided to show the potential application of HDDCS in the digital world. The basic advantage and practical benefits of the proposed approach is that the chaos satisfying Devaney's definition can be generated in the real digital devices by means of the chaos generation strategy controlled by random sequences.

Finally, in the last chapter, software implementations of the proposal are deeply investigated using a large collection of statistical tests. We show in various contexts that chaotic iterations can improve the statistical behavior of inputted pseudorandom number algorithms, and we propose various posttreatments to do so. This book ends by a conclusion, an appendix that details some well-known pseudorandom number generators that are used here, and various references at the interface of chaos and randomness.

References

1. E.N. Lorent, Deterministic non-periodic flow. J. Atmos. Sci. **20**, 130–141 (1963)
2. L.O. Chua, M. Komuro, T. Matsumoto, The double scroll family. IEEE Trans. Circuits Syst. I **33**(11), 1072–1118 (1986)
3. G. Chen, T. Ueta, Yet another chaotic attractor. Int. J. Bifurc. Chaos **9**(07), 1465–1466 (1999)
4. S. Yu, G. Chen, Chaotifying continuous-time nonlinear autonomous systems. Int. J. Bifurc. Chaos **22**(9), 1250232-1 (2012). art. no. 1250232
5. R.M. May, Simple mathematical models with very complicated dynamics. Nature **261**(5560), 459–467 (1976)
6. F. Zheng, X. Tian, J. Song, X. LI, Pseudo-random sequence generator based on the generalized Henon map. J. China Univ. Posts Telecommun. **15**(3), 64–68 (2008)
7. G. Chen, D. Lai, Making a dynamical system chaotic: feedback control of Lyapunov exponents for discrete-time dynamical systems. IEEE Trans. Circuits Syst. I **44**(3), 250–253 (1997)
8. Z. Lin, S. Yu, J. Lu, S. Cai, G. Chen, Design and ARM-embedded implementation of a chaotic map-based real-time secure video communication system. IEEE Trans. Circuits Syst. Video Technol. **25**(7), 1203–1216 (2015)
9. S. Vaidyanathan, A.T. Azar, Dynamic analysis, adaptive feedback control and synchronization of an eight-term 3-D novel chaotic system with three quadratic nonlinearities, *Advances in Chaos Theory and Intelligent Control* (Springer International Publishing, Cham, 2016), pp. 155–178

10. V.T. Pham, S. Jafari, C. Volos, A. Giakoumis, S. Vaidyanathan, T. Kapitaniak, A chaotic system with equilibria located on the rounded square loop and its circuit implementation. IEEE Trans. Circuits Syst. II: Express Briefs **63**(9), 878–882 (2016)

11. C. Shen, S. Yu, J. Lü, G. Chen, A systematic methodology for constructing hyperchaotic systems with multiple positive Lyapunov exponents and circuit implementation. IEEE Trans. Circuits Syst. I **61**(3), 854–864 (2014)

12. G. Alvarez, S. Li, Some basic cryptographic requirements for chaos-based cryptosystems. Int. J. Bifurc. Chaos **16**(8), 2129–2151 (2006)

13. C. Shen, S. Yu, J. Lü, G. Chen, Designing hyperchaotic systems with any desired number of positive Lyapunov exponents via a simple model. IEEE Trans. Circuits Syst. I **61**(8), 2380–2389 (2014)

14. F.Y. Shih, *Digital Watermarking and Steganography: Fundamentals and Techniques* (CRC Press, Boca Raton, 2017)

15. T. Kohda, A. Tsuneda, Statistics of chaotic binary sequences. IEEE Trans. Inf. Theory **43**(1), 104–112 (1997)

16. G.D. VanWiggeren, R. Roy, Communication with chaotic lasers. Science **279**(5354), 1198–1200 (1998)

17. S. Tan, J. Lü, D. Hill, Towards a theoretical framework for analysis and intervention of random drift on general networks. IEEE Trans. Autom. Contr. **60**(2), 576–581 (2015)

18. K. Liu, H. Zhu, J. Lü, Bridging the gap between transmission noise and sampled data for robust consensus of multi-agent systems. IEEE Trans. Circuits Syst. I **62**(7), 1836–1844 (2015)

19. K. Liu, L. Wu, J. Lü, H. Zhu, Finite-time adaptive consensus of a class of multi-agent systems. Sci. China Technol. Sci. **59**(1), 22–32 (2016)

20. G. Chen, Y. Mao, C.K. Chui, A symmetric image encryption scheme based on 3d chaotic cat maps. Chaos Solitons Fractals **21**(3), 749–761 (2004)

21. A. Argyris, D. Syvridis, L. Larger, V. Annovazzi-Lodi, P. Colet, I. Fischer, J. Garcia-Ojalvo, C.R. Mirasso, L. Pesquera, K.A. Shore, Chaos-based communications at high bit rates using commercial fiber-optic links. Nature **438**, 343–346 (2005)

22. X. Li, C. Li, I. Lee, Chaotic image encryption using pseudo-random masks and pixel mapping. Signal Process. **125**(Supplement C), 48–63 (2016)

23. Q. Jiang, F. Wei, S. Fu, J. Ma, G. Li, A. Alelaiwi, Robust extended chaotic maps-based three-factor authentication scheme preserving biometric template privacy. Nonlinear Dyn. **83**(4), 2085–2101 (2016)

24. F. Dachselt, W. Schwarz, Chaos and cryptography. IEEE Trans. Circuits Syst. I **48**(12), 1498–1509 (2001)

25. E. Tlelo-Cuautle, J. Rangel-Magdaleno, L.G. de la Fraga, *Engineering Applications of FP-GAs: Chaotic Systems, Artificial Neural Networks, Random Number Generators, and Secure Communication Systems* (Springer, Berlin, 2016)

26. B. Muthuswamy, S. Banerjee, *A Route to Chaos Using FPGAs: Volume I: Experimental Observations* (Springer, Berlin, 2015)

27. S. Li, G. Chen, X. Mou, On the dynamical degradation of digital piecewise linear chaotic maps. Int. J. Bifurc. Chaos **15**(10), 3119–3151 (2005)

28. H.G. Schuster, W. Just, *Deterministic Chaos: An Introduction* (Wiley, New York, 2006)

29. C. Beck, G. Roepstorff, Effects of phase space discretization on the long-time behavior of dynamical systems. Phys. D **25**(1), 173–180 (1987)

30. M. Blank, Pathologies generated by round-off in dynamical systems. Phys. D **78**(1), 93–114 (1994)

31. P.M. Binder, R.V. Jensen, Simulating chaotic behavior with finite-state machines. Phys. Rev. A **34**(5), 4460–4463 (1986)

32. S. Li, X. Mou, Y. Cai, Z. Ji, J. Zhang, On the security of a chaotic encryption scheme: Problems with computerized chaos in finite computing precision. Comput. Phys. Commun. **153**(1), 52–58 (2003)

33. Y. Deng, H. Hu, W. Xiong, N.N. Xiong, L. Liu, Analysis and design of digital chaotic systems with desirable performance via feedback control. IEEE Trans. Syst. Man Cybernet.: Syst. **45**(8), 1187–1200 (2015)

34. M. Stability, T. Zourntos, D.A. Johns, Fundamental theory and applications. IEEE Trans. Circuits Syst. **I**(49), 41–53 (2002)
35. C. Li, Cracking a hierarchical chaotic image encryption algorithm based on permutation. Signal Process. **118**, 203–210 (2016)
36. S. Čelikovský, V. Lynnyk, Message embedded chaotic masking synchronization scheme based on the generalized Lorenz system and its security analysis. Int. J. Bifurc. Chaos **26**(08) (2016). p.Art.no. 1650140
37. Z. Galias, The dangers of rounding errors for simulations and analysis of nonlinear circuits and systems-and how to avoid them. IEEE Circuits Syst. Mag. **13**(3), 35–52 (2013)
38. Y. Deng, H. Hu, N. Xiong, W. Xiong, L. Liu, A general hybrid model for chaos robust synchronization and degradation reduction. Inf. Sci. **305**, 146–164 (2015)
39. J. Cernak, Digital generators of chaos. Phys. Lett. A **214**(3–4), 151–160 (1996)
40. T. Sang, R. Wang, Y. Yan, Perturbance-based algorithm to expand cycle length of chaotic key stream. Electron. Lett. **34**(9), 873–874 (1998)
41. C.-Y. Li, Y.-H. Chen, T.-Y. Chang, L.-Y. Deng, K. To, Period extension and randomness enhancement using high-throughput reseeding-mixing prng. IEEE Trans. Very Large Scale Integrat. Syst. **20**(2), 385–389 (2012)
42. T. Lin, L. Chua, On chaos of digital filters in the real world. IEEE Trans. Circuits Syst. I **38**(5), 557–558 (1991)
43. H. Zhou, X. Ling, Realizing finite precision chaotic systems via perturbation of m-sequences. Acta Electron. Sin. **25**(7), 95–97 (1997). (in Chinese)
44. L. Liu, J. Lin, S. Miao, B. Liu, A double perturbation method for reducing dynamical degradation of the digital baker map. Int. J. Bifurc. Chaos **27**(07) (2017). p.Art.no.1750103
45. W. Wolff, B. Huberman, Transients and asymptotics in granular phase space. Zeitschrift für Phys. B Condens. Matter **63**(3), 397–405 (1986)
46. N. Nagaraj, M.C. Shastry, P.G. Vaidya, Increasing average period lengths by switching of robust chaos maps in infinite precision. Eur. Phys. J. Spec. Top. **165**(1), 73–83 (2008)
47. Y. Wu, Y. Zhou, L. Bao, Discrete wheel-switching chaotic system and applications. IEEE Trans. Circuits Syst. I **61**(12), 3469–3477 (2014)
48. C. Li, T. Xie, Q. Liu, G. Cheng, Cryptanalyzing image encryption using chaotic logistic map. Nonlinear Dyn. **78**(2), 1545–1551 (2014)
49. C. Guyeux, J. Bahi, Hash functions using chaotic iterations. J. Algorithms Comput. Technol. **4**(2), 167–182 (2010)
50. R.L. Devaney, *An Introduction to Chaotic Dynamical Systems* (Westview Press, Boulder, 2003)
51. J. Bahi, J.F. Couchot, C. Guyeux, Q.X. Wang, Class of trustworthy pseudo random number generators, in *IEEE International Conference on Evolving Internet, Luxembourg, June 2011*, pp. 72–77
52. J.M. Bahi, X. Fang, C. Guyeux, Q. Wang, Evaluating quality of chaotic pseudo-random generators: Application to information hiding. Int. J. Adv. Secur. **4**(1–2), 118–130 (2011)
53. J.M. Bahi, X. Fang, C. Guyeux, Q. Wang, Suitability of chaotic iterations schemes using xorshift for security applications. J. Netw. Comput. Appl. **37**, 282–292 (2014)
54. J. Banks, J. Brooks, G. Cairns, G. Davis, P. Stacey, On devaney's definition of chaos. Am. Math. Mon. **99**(4), 332–334 (1992)
55. T.Y. Li, J.A. Yorke, Period three implies chaos. Am. Math. Mon. **82**(10), 985–992 (1975)
56. R. Simard, U.D. Montréal, Testu01: a software library in ansi c for empirical testing of random number generators. ACM Trans. Math. Softw. **33**(4), 22 (2007)
57. A. Rukhin, J. Soto, J. Nechvatal et al., A statistical test suite for random and pseudorandom number generators for cryptographic applications, NIST Special Publication 800-22rev1a (2010), http://csrc.nist.gov/groups/ST/toolkit/rng/documentation_software.html
58. G. Marsaglia, Diehard battery of tests of randomness, Florida State University, 1995
59. Q. Wang, S. Yu, C. Guyeux, J. Bahi, X. Fang, Theoretical design and circuit implementation of integer domain chaotic systems. Int. J. Bifurc. Chaos **24**(10), 1450128-1 (2014). p. Art. no. 1450128
60. A. Hajimiri, T. Lee, A general theory of phase noise in electrical oscillators. IEEE J. Solid-State Circuits **33**(2), 179–194 (1998)

Chapter 2
Integer Domain Chaotic Systems (IDCS)

2.1 Description of IDCS

In this section, we first introduce the basic concept of integer domain chaotic systems (IDCS), and compare them to real domain chaotic systems.

2.1.1 Real Domain Chaotic Systems (RDCS)

In traditional RDCS studies, the general form of the iterative equation is:

$$x^0 \in \mathbb{R}, \text{ and} \forall n \in \mathbb{N}, x^n = f(x^{n-1}),$$

where $f : \mathbb{R} \to \mathbb{R}$ is the iterative equation, and x^{n-1} and x^n are the $n-1$th and nth iteration, respectively. Note that x^{n-1} and x^n are real numbers, which are represented in binary form as

$$\begin{cases} x^{n-1} = x_{i_1} x_{i_2} \dots x_{i_M} . x_{j_1} x_{j_2} \dots x_{j_N} \dots, \\ x^n = x_{k_1} x_{k_2} \dots x_{k_L} . x_{l_1} x_{l_2} \dots x_{l_P} \dots, \end{cases} \tag{2.1}$$

where $x_{i_1}, x_{i_2}, \dots, x_{i_M} \in \{0, 1\}$ and $x_{j_1}, x_{j_2}, \dots, x_{j_N}, \dots \in \{0, 1\}$ are the integer and fractional parts for x^{n-1}. Similarly, $x_{k_1}, x_{k_2}, \dots, x_{k_L} \in \{0, 1\}$ and $x_{l_1}, x_{l_2}, \dots, x_{l_P}, \dots \in \{0, 1\}$ are, respectively, the integer and fractional parts for x^n.

The main features of discrete-time RDCS are that all the bits in x^{n-1} will be updated by iterative equation f at each operation (iteration). Likewise, all the bits in x^n will be updated by iterative equation f at each operation (iteration).

Parts of this chapter were reproduced with permission from [1] ©World Scientific Publishing Co Pte Ltd 2014, [2] ©Chinese Physical Society and IOP Publishing Ltd 2015, and [3] ©IEEE 2016.

2.1.2 IDCS

The main ideas of CI (chaotic iteration) systems are summarized in the following.

Let $N \in \{1, 2, \ldots\}$ be a positive integer, $\mathbb{B} = \{1, 0\}$ denotes the set of binary numbers, and $\mathbb{B}^N$ is the set of binary vectors of size N. For any $n = 0, 1, 2, \ldots, x^n$ is represented by using N bits in base-2: $x^0 = (x^0_{N-1} x^0_{N-2} \ldots x^0_0) \in \mathbb{B}^N$ is the initial condition, whereas $x^{n-1} = (x^{n-1}_{N-1} x^{n-1}_{N-2} \ldots x^{n-1}_0) \in \mathbb{B}^N$ and $x^n = (x^n_{N-1} x^n_{N-2} \ldots x^n_0) \in \mathbb{B}^N$ denote the $n - 1$th and nth iterations, respectively. In CI systems, the iterative equation is defined as

$$x^n_i = \begin{cases} x^{n-1}_i & \text{if } i \neq s^n, \\ f(x^{n-1})_i & \text{if } i = s^n, \end{cases} \tag{2.2}$$

where $i = 0, 1, 2, \ldots, N - 1$, $n = 1, 2, \ldots$, and $s = (s^1 s^2 \ldots s^n \ldots)$ is a one-sided infinite sequence of integers bounded by $N - 1$: $\forall n \in \mathbb{N}$, $s^n \in \{0, 1, 2, \ldots, N - 1\}$. Additionally, the iterative function f is usually the vectorial Boolean negation, given by

$$f(x^{n-1}) = (\overline{x^{n-1}_{N-1}} \, \overline{x^{n-1}_{N-2}} \ldots \overline{x^{n-1}_i} \ldots \overline{x^{n-1}_0}), \tag{2.3}$$

and the following notation is used,

$$f(x^{n-1})_{i=s^n} = (\overline{x^{n-1}_{N-1}} \, \overline{x^{n-1}_{N-2}} \ldots \overline{x^{n-1}_i} \ldots \overline{x^{n-1}_0})_{i=s^n} = \overline{x^{n-1}_{i=s^n}}, \tag{2.4}$$

that is, $f(x^{n-1})_{i=s^n}$ is the ith component of $f(x^{n-1})$. Let us finally remark that, in IDCS, the one-sided infinite sequence of integers $s = (s^1 s^2 \ldots s^n \ldots)$ is usually called a chaotic strategy.

For the one-sided infinite sequence of integers $s^n (n = 1, 2, \ldots) \in \{0, 1, 2, \ldots, N - 1\}$ in IDCS, one of the options is:

$$s^n = R(\bmod N). \tag{2.5}$$

In the above equation, $R = \{0, 1, \ldots, 2^k - 1\}(k = 1, 2, \ldots)$ is a uniformly distributed random integer (generated, for instance, by a random number generator), and $R > N$ is normally taken. For example, $R = 0, 1, \ldots, 2^{32} - 1$ can be generated by the LCG (linear congruential generator) and $N = 4$ can be chosen. Because 2^{32} can be divided by 4, $s^n (n = 1, 2, \ldots) \in \{0, 1, 2, \ldots, N - 1\}$ is a uniformly distributed random integer after the modulo operation. But if 2^k cannot be divided by N, s^n is not a strictly uniformly distributed random integer, especially when R is close to N, although we have uniform distribution of R.

Note that, in Eq. (2.2), if the output is x^n at each iteration, then it is proven to satisfy Devaney's definition of chaos (see below), but its safety is not enough for cryptography purposes. The main reason is that only one bit is modified between two successive iterations, and the remaining bits are not changed. In order to solve this problem, only x at every m iterations is outputted. As we iterate m times between two outputs, each bit of x has been updated m/N times on average. In this way, it is possible to change a random number of bits between two adjacent outputs, so the number of the changed bits is not fixed.

For example, the chaotic strategy can be:

$$m = 3N + RR(\text{mod}\,2), \tag{2.6}$$

where RR is generated by a RNG. This strategy has been called the OldCI algorithm [4].

Let x_k, x_j be two binary digits, the corresponding distance is

$$\delta(x_j, x_k) = \begin{cases} 1 \text{ if } x_j \neq x_k, \\ 0 \text{ if } x_j = x_k. \end{cases} \tag{2.7}$$

Using the same notations as above, we define the binary iterative equation as

$$F_f(k, x)_j = x_j \cdot \delta(k, j) + f(x)_j \cdot \overline{\delta(k, j)}, \tag{2.8}$$

where $j \in \{0, 1, 2, \ldots, N - 1\}$, and k is usually a term of chaotic strategy s whereas f is often the binary vectorial negation recalled previously. With these choices, and according to Eq. (2.8), a more specific formula can be obtained:

$$F_f(k, x) = (x_{N-1}, x_{N-2}, \ldots, x_{k+1}, \overline{x_k}, x_{k-1}, \ldots, x_1, x_0).$$

Let $E = (s, x)$ be a couple constituted by a chaotic strategy and a Boolean vector; that is, $E = (s, x) \in \mathscr{E} = \{0, 1, 2, \ldots, N - 1\}^\infty \times \mathbb{B}^N$. Define function G_f as

$$G_f(E) = G_f((s, x)) = (\sigma(s), F_f(i(s), x)), \tag{2.9}$$

where $i(s) = s^1$ and $\sigma^k(s) = \underbrace{\sigma \circ \sigma \circ \ldots \circ \sigma(s)}_{k}, k = 1, 2, \ldots,$ is the result of shifting k integers in the one-sided infinite sequence $s = (s^1 s^2 \ldots s^n \ldots)$ to the left. In other words,

$$\sigma^k(s) = s^{k+1} s^{k+2} \ldots s^n \ldots (k = 1, 2, \ldots).$$

With all this material, IDCS is defined as

$$E^0 \in \mathscr{E} \text{ and } \forall k \in \mathbb{N}, E^{k+1} = G_f(E^k).$$

Consider now two real numbers a and b, lower than 1, that are represented in radix-r format as

$$\begin{cases} a = 0.a_1 a_2 a_3 \ldots a_n \ldots = \sum_{k=1}^{\infty} \frac{a_k}{r^k}, \\ b = 0.b_1 b_2 b_3 \ldots b_n \ldots = \sum_{k=1}^{\infty} \frac{b_k}{r^k}, \end{cases} \tag{2.10}$$

where $a_k, b_k \in \{0, 1, 2, \ldots, r-1\}$. Then, the distance between a and b is given by:

$$d(a, b) = \sum_{k=1}^{\infty} \frac{|a_k - b_k|}{r^k}. \qquad (2.11)$$

The above formula can be generalized to calculate the distance between two one-sided infinite sequences of symbols without loss of generality. These remarks lead to the definition of a new distance on the set $\mathscr{E}$, which is defined by:

$$d((s, x), (\hat{s}, \hat{x})) = d_s(s, \hat{s}) + d_x(x, \hat{x}),$$

where $s = (s^1 s^2 \ldots s^n \ldots)$ and $\hat{s} = (\hat{s}^1 \hat{s}^2 \ldots \hat{s}^n \ldots)$ are one-sided infinite sequences of integers, and x and $\hat{x}$ are binary digits of N bits. More precisely, and in agreement with Eq. (2.11), the distance between s and $\hat{s}$ is:

$$d_s(s, \hat{s}) = \sum_{k=1}^{\infty} \frac{|s^k - \hat{s}^k|}{N^k} \in [0, 1], \qquad (2.12)$$

where $\forall k \in \mathbb{N}, s^k, \hat{s}^k \in \{0, 1, 2, \ldots, N-1\}$. Finally, following the 1-norm distance, the distance between x and $\hat{x}$ is:

$$d_x(x, \hat{x}) = \sum_{k=0}^{N-1} \delta(x_k, \hat{x}_k) \in \{0, 1, 2, \ldots, N\}. \qquad (2.13)$$

Remark that d is a distance on the set $\mathscr{E}$, as it is defined as the sum of two distances.

Before investigating the chaotic properties of IDCS, we have to prove that G_f is continuous on the metric space $(\mathscr{E}, d)$. In order to do so the following lemma is first established.

Lemma 2.1 *Let $s = (s^1 s^2 \ldots s^n \ldots)$ and $\hat{s} = (\hat{s}^1 \hat{s}^2 \ldots \hat{s}^n \ldots)$, where*

$$s^k, \hat{s}^k \in \{0, 1, 2, 3, \ldots, N-1\}$$

for $k = 1, 2, \ldots$. If $s^i = \hat{s}^i$ for $i = 1, 2, \ldots, n$, then $d(s, \hat{s}) \leq \frac{1}{N^n}$. Conversely, if $d(s, \hat{s}) \leq \frac{1}{N^n}$, then $s^i = \hat{s}^i$ for $i = 1, 2, \ldots, n$.

Proof If $s^i = \hat{s}^i$ ($i = 1, 2, \ldots, n$), then

$$d(s, \hat{s}) = \sum_{i=1}^{n} \frac{|s^i - \hat{s}^i|}{N^i} + \sum_{i=n+1}^{\infty} \frac{|s^i - \hat{s}^i|}{N^i} = \sum_{i=n+1}^{\infty} \frac{|s^i - \hat{s}^i|}{N^i}$$

$$\leq \sum_{i=n+1}^{\infty} \frac{N-1}{N^i} = (N-1)\frac{\frac{1}{N^{n+1}}}{1 - \frac{1}{N}} = \frac{1}{N^n}.$$

Conversely, and due to the definition of the proposed distance: for any $m \leqslant n$, if $s^m \neq \hat{s}^m$, then $d(s, \hat{s}) \geqslant \frac{1}{N^n}$. The contraposition is the desired result: if $d(s, \hat{s}) \leqslant \frac{1}{N^n}$, then $s^i = \hat{s}^i$ ($i = 1, 2, \ldots, n$).

To prove that chaotic iterations are an example of chaos, we must first set that G_f is continuous on the metric space $(\mathcal{E}, d)$.

Theorem 2.1 G_f is a continuous function.

Proof A continuous function is a function for which, intuitively, "small" changes in the input result in "small" changes in the output. Let $((s, x)^n)_{n \in \mathbb{N}}$ be a sequence of the phase space $\mathcal{E}$, which converges to $(\hat{s}, \hat{x})$. We prove that $G_f^n(s, x)_{n \in \mathbb{N}}$ converges to $G_f(\hat{s}, \hat{x})$. In mathematical notation, $\forall((s, x)^n)_{n \in \mathbb{N}} \in \mathcal{E}^{\mathbb{N}} : \lim_{n \to \infty} (s, x)^n = (\hat{s}, \hat{x}) \Rightarrow \lim_{n \to \infty} G_f^n(s, x) = G_f(\hat{s}, \hat{x})$.

1. $\lim_{n \to \infty} (s, x)^n = (\hat{s}, \hat{x}) \Rightarrow \forall \varepsilon > 0, d((s, x)^n, (\hat{s}, \hat{x})) < \varepsilon$ for large n.
   Thus, without loss of generality, we assume that $\varepsilon < 1$.
2. If $x^n \neq \hat{x}$, then $d_x(x^n, \hat{x}) \geqslant 1$, and so $d((s, x)^n, (\hat{s}, \hat{x})) = d_s(s^n, \hat{s}) + d_x(x^n, \hat{x}) > \varepsilon$.
   Thus there $\exists n_0 \in \mathbb{N}, d_x(x^n, \hat{x}) = 0$ for any $n \geqslant n_0$.
3. As $d((s, x)^n, (\hat{s}, \hat{x})) < \varepsilon$ leads to

$$d((s, x)^n, (\hat{s}, \hat{x})) = d_s(s^n, \hat{s}) + d_x(x^n, \hat{x}) = d_s(s^n, \hat{s}) < \varepsilon,$$

according to the previous Lemma 2.1; if the first k_0 elements of s^n and $\hat{s}$ are the same, then $d_s(s^n, \hat{s}) < N^{-k_0} < \varepsilon$. For instance, $k_0 = \lfloor -log_N \varepsilon \rfloor + 1$ is convenient. Thus there $\exists n_1 = k_0 \in \mathbb{N}, d_s(s^n, \hat{s}) < \varepsilon$ for any $n \geqslant n_1$.
4. According to Eq. (2.9), the corresponding $G_f^n(s, x)$ and $G_f(\hat{s}, \hat{x})$ can be obtained:

$$G_f^n(s, x) = (\sigma(s^n), F_f(i(s^n), x^n)),$$

$$G_f(\hat{s}, \hat{x}) = (\sigma(\hat{s}), F_f(i(\hat{s}), \hat{x})).$$

For $n \geqslant max(n_0, n_1)$, the first k_0 elements of s^n and $\hat{s}$ are the same and $x^n = \hat{x}$, therefore

$$i(s^n) = i(\hat{s}).$$

Then,

$$F_f(i(s^n), x^n) = F_f(i(\hat{s}), \hat{x}).$$

$\sigma(s)$ is the result of shifting one integer in the one-sided infinite sequence to the left. Therefore the first $k_0 - 1$ elements of $\sigma((s)_n)$ and $\sigma(\hat{s})$ are still the same. Thus

$$d(G_f((s, x)^n), G_f(\hat{s}, \hat{x})) = d_s(\sigma(s^n), \sigma(\hat{s})) + d_x(F_f(i(s^n), x^n), F_f(i(\hat{s}), \hat{x}))$$
$$= d_s(\sigma(s^n), \sigma(\hat{s})) < N^{-(k_0-1)},$$

which makes

$$\lim_{n\to\infty} G_f^n(s, x) = G_f(\hat{s}, \hat{x})$$

true.

In conclusion, G_f is consequently continuous.

2.2 Proof of Chaos for IDCS

In this section, the chaotic behavior of IDCS is proven according to Devaney's definition.

2.2.1 Dense Periodic Points

Theorem 2.2 *The periodic points of G_f are dense in $(\mathscr{E}, d)$.*

Proof We want to show that, for any given $\varepsilon > 0$, a periodic point $(\tilde{s}, \tilde{x}) \in \mathscr{E}$ can always be found within range ε of any point $(\hat{s}, \hat{x})$ in $(\mathscr{E}, d)$.

1. Without loss of generality, we assume that the given $\varepsilon < 1$ and that the general form of $(\hat{s}, \hat{x})$ is

$$(\hat{s}, \hat{x}) = ((\hat{s}^1 \hat{s}^2 \dots \hat{s}^{k_0} \dots \hat{s}^n \dots), \hat{x}) \in \mathscr{E}.$$

2. If $\tilde{x} \neq \hat{x}$, then $d_x(\tilde{x}, \hat{x}) \geqslant 1$, thus $d((\hat{s}, \hat{x}), (\tilde{s}, \tilde{x})) > 1$. Thus $\tilde{x} = \hat{x}$.
3. If the first k_0 elements of $\hat{s}$ and $\tilde{s}$ are the same, then $d_s(\hat{s}, \tilde{s}) < N^{-k_0}$ according to the previous Lemma 2.1. Thus, $\forall \varepsilon < 1$, an integer k_0 can always be found making the relation $d_s(\hat{s}, \tilde{s}) < N^{-k_0} < \varepsilon$ true. For instance, $k_0 = \lfloor -log_N \varepsilon \rfloor + 1$ is convenient.
4. If after the k_0th iteration, we have

$$(G_f^{k_0}(\tilde{s}, \tilde{x}))_x = \tilde{x} = \hat{x},$$

then a cycle point $(\tilde{s}, \tilde{x}) = ((\hat{s}^1 \hat{s}^2 \dots \hat{s}^{k_0} \hat{s}^1 \hat{s}^2 \dots \hat{s}^{k_0} \dots), \hat{x}) \in \mathscr{E}$ is found that satisfies

$$(\tilde{s}, \tilde{x}) = G_f^{k_0}(\tilde{s}, \tilde{x}),$$

making

$$d((\hat{s}, \hat{x}), (\tilde{s}, \tilde{x})) = d_s(\hat{s}, \tilde{s}) + d_x(\hat{x}, \tilde{x}) = d_s(\hat{s}, \tilde{s}) < \varepsilon$$

true.

5. If after the k_0th iteration, we have

$$(G_f^{k_0}(\tilde{s}, \tilde{x}))_x \neq \tilde{x};$$

then, without loss of generality, we can assume that there are $i_0 \, (\leqslant N)$ different bits between $\hat{x}$ and $\tilde{x}$. These i_0 bits are numbered $j^1 < j^2 < \ldots < j^{i_0}$, respectively. To obtain that, after another i_0 iterations, the following condition is met

$$\tilde{x} = \hat{x} = (G_f^{k_0+i_0}(\tilde{s}, \tilde{x}))_x,$$

we must set:

$$\begin{cases} \tilde{s}^{k_0+1} = j^1, \\ \tilde{s}^{k_0+2} = j^2, \\ \quad\vdots \\ \tilde{s}^{k_0+i_0} = j^{i_0}. \end{cases} \tag{2.14}$$

Then, within the range ε of the point $(\hat{s}, \hat{x})$, one can find the periodic point:

$$\begin{aligned} (\tilde{s}, \tilde{x}) &= ((\hat{s}^1 \hat{s}^2 \ldots \hat{s}^{k_0} j^1 j^2 \ldots j^{i_0} \hat{s}^1 \hat{s}^2 \ldots \hat{s}^{k_0} j^1 j^2 \ldots j^{i_0} \ldots), \hat{x}) \\ &= G_f^{k_0+i_0}(\tilde{s}, \tilde{x}) \in \mathscr{E} \end{aligned}$$

making

$$d((\hat{s}, \hat{x}), (\tilde{s}, \tilde{x})) = d_s(\hat{s}, \tilde{s}) + d_x(\hat{x}, \tilde{x}) = d_s(\hat{s}, \tilde{s}) < \varepsilon$$

true.

In summary, the periodic points of G_f are dense in $(\mathscr{E}, d)$.

2.2.2 Transitive Property

Theorem 2.3 G_f *is a transitive map in* $(\mathscr{E}, d)$.

Proof The so-called topological transitivity specifically refers to that: for any non-empty open sets U_A and U_B in $(\mathscr{E}, d)$, there is always $n_0 > 0$ which makes $G_f^{n_0}(U_A) \cap U_B \neq \emptyset$.

Consider now two nonempty open sets U_A and U_B, and $(s_A, x_A) \in U_A$, $(s_B, x_B) \in U_B$. U_A and U_B are open, and we take place in a metric space, thus there exist real numbers $r_A > 0$ and $r_B > 0$ such that the open ball $\mathscr{B}_A$ with center (s_A, x_A) and radius r_A is inside U_A (resp., the open ball $\mathscr{B}_B$ with center (s_B, x_B) and radius r_B is inside U_B). Without loss of generality, we can suppose that $r_A < 1$.

1. We introduce the following notations.

$$(s_A, x_A) = ((s_A^1 s_A^2 \ldots s_A^{n_0} \ldots s_A^n \ldots), x_A) \in U_A \subseteq \mathscr{E}$$

and

$$(s_B, x_B) = ((s_B^1 s_B^2 \ldots s_B^n \ldots), x_B) \in U_B \subseteq \mathscr{E}.$$

2. Let $(\tilde{s}, \tilde{x}) \in U_A$. If $\tilde{x} \neq x_A$, then $d_x(\tilde{x}, x_A) \geq 1$, and therefore $d((s_A, x_A), (\tilde{s}, \tilde{x})) > 1$. Consequently, if $(\tilde{s}, \tilde{x}) \in \mathscr{B}_A$, then $d((s_A, x_A), (\tilde{s}, \tilde{x})) < r_A < 1$, and thus $\tilde{x} = x_A$.

3. If we demand that the first k_0 elements of $\tilde{s}$ are equal to those from s_A, then we obtain $d_s(s_A, \tilde{s}) < N^{-k_0}$. And for the given r_A, an integer k_0 and a sequence $\tilde{s}$ can always be found to achieve $d_s(s_A, \tilde{s}) < N^{-k_0} < r_A$ (for instance, $k_0 = \lfloor -log_N r_A \rfloor + 1$).

4. If after k_0 iterations, the following condition is satisfied,

$$(G_f^{k_0}(\tilde{s}, \tilde{x}))_x = x_B,$$

then $n_0 = k_0$ and $(\tilde{s}, \tilde{x}) = ((s_A^1 s_A^2 \ldots s_A^{n_0} s_B^1 s_B^2 \ldots s_B^n \ldots), x_A) \in U_A$ have been found that satisfy:

$$G_f^{n_0}(\tilde{s}, \tilde{x}) = (s_B, x_B) \in G_f^{n_0}(U_A) \cap U_B,$$

making

$$G_f^{n_0}(U_A) \cap U_B \neq \varnothing$$

true.

5. If, after the k_0th iteration,

$$(G_f^{k_0}(\tilde{s}, \tilde{x}))_x \neq x_B,$$

then, without loss of generality, we can assume that there are i_0 ($\leq N$) different bits between x_B and the Boolean vector of $(G_f^{k_0}(\tilde{s}, \tilde{x}))_x$. Once again, these i_0 bits are numbered $j^1 < j^2 < \ldots < j^{i_0}$, respectively. Define now,

$$\begin{cases} \tilde{s}^{k_0+1} = j^1, \\ \tilde{s}^{k_0+2} = j^2, \\ \quad \vdots \\ \tilde{s}^{k_0+i_0} = j^{i_0}, \end{cases} \tag{2.15}$$

therefore the point $(\tilde{s}, \tilde{x}) = ((s_A^1 s_A^2 \ldots s_A^{k_0} j^1 j^2 \ldots j^{i_0} s_B^1 s_B^2 \ldots s_B^n \ldots), x_A) \in U_A$ satisfies

$$G_f^{n_0}(\tilde{s}, \tilde{x}) = (s_B, x_B) \in G_f^{n_0}(U_A) \cap U_B$$

with $n_0 = k_0 + i_0$, making the claim

$$G_f^{n_0}(U_A) \cap U_B \neq \varnothing$$

true.

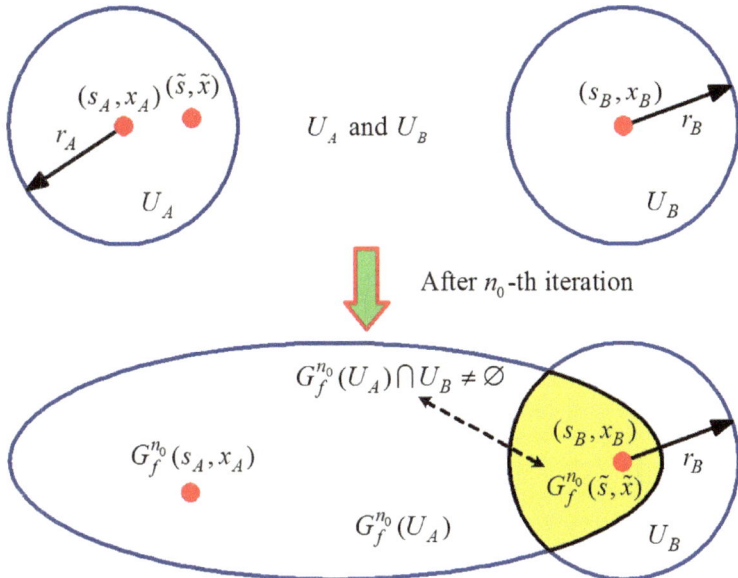

Fig. 2.1 The schematic diagram of transitivity in $(\mathscr{E}, d)$. (©World Scientific Publishing Co Pte Ltd 2014, Reproduced with permission from [1])

In summary, $(\mathscr{E}, d)$ is transitive, as shown in Fig. 2.1.

Due to Definition 1.1 and Theorem 1.1 in Chap. 1, and as we stated the density of periodic points and the transitivity property, we can claim that IDCS is chaotic in the sense of Devaney.

2.2.3 Further Investigations of the Chaotic Behavior of IDCS

Thus the set $\mathscr{C}$ of functions $f : \mathbb{B}^N \longrightarrow \mathbb{B}^N$ making the IDCS a case of chaos according to Devaney, is a nonempty set. We have further established that:

Theorem 2.1 $\forall f \in \mathscr{C}$, *the number of periodic points of G_f is infinitely countable, G_f is strongly transitive, and is chaotic according to Knudsen. It is thus undecomposable, unstable, and chaotic as defined by Wiggins.*

Additionally, we have obtained the following results for the vectorial negation.

Theorem 2.2 G_{f_0} *is topologically mixing, expansive (with a constant equal to 1), chaotic as defined by Li and Yorke, and has a topological entropy equal to $ln(N)$.*

2.2.4   Relationship Between Iterative Input and Output

As is well known, for a dynamical system to display a chaotic behavior, it must be either nonlinear or infinite-dimensional. If the relationship between the input and the output of the finite-dimensional system is linear, as shown in Fig. 2.2, the system is not a chaotic one.

When the relationship between the input and the output of the finite-dimensional system is nonlinear, the system can behave chaotically. Therefore, it is necessary to discuss the relationship between the input and the output of IDCS. To illustrate this fact, let us take the following analysis and discussion ($N = 4$, $x^0 = 0$ and the vectorial Boolean negation as iterative equation) as an example.

Let $x^n = x_3^n x_2^n x_1^n x_0^n$, $x^{n+1} = x_3^{n+1} x_2^{n+1} x_1^{n+1} x_0^{n+1}$, and $x^0 = x_3^0 x_2^0 x_1^0 x_0^0$ be the nth, $n + 1$th, and initial iteration values, respectively, and the negation with one bit is run in this loop: $x_0^n \rightarrow x_1^{n+1} \rightarrow x_2^{n+2} \rightarrow x_3^{n+3}$. That is, the logical value of the 0th bit of x is reversed at the first iteration, then the logical value of the 1st bit of x is reversed at the second iteration, followed by the reverse of the 2nd bit of x, the 3rd bit of x is reversed, the logical value of the 0th bit of x is reversed again, and so on. The result of the iterations is as follows.

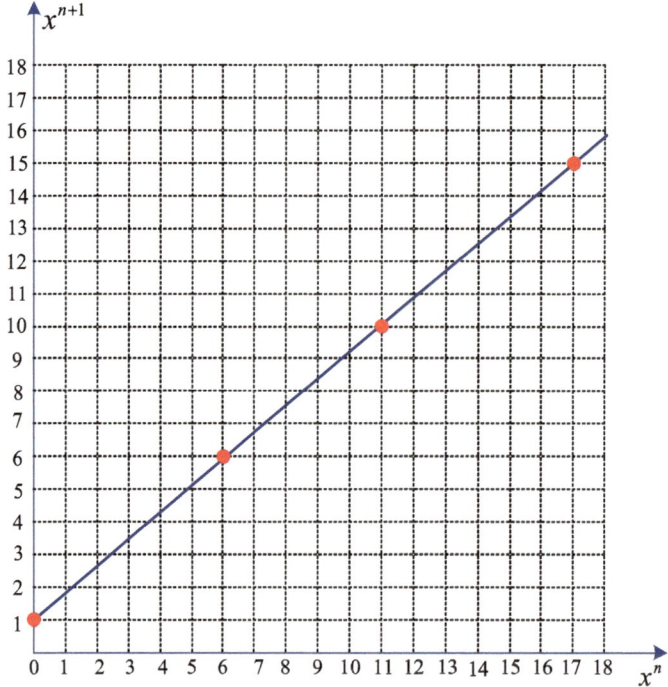

Fig. 2.2 The linearization between the input and the output of the system

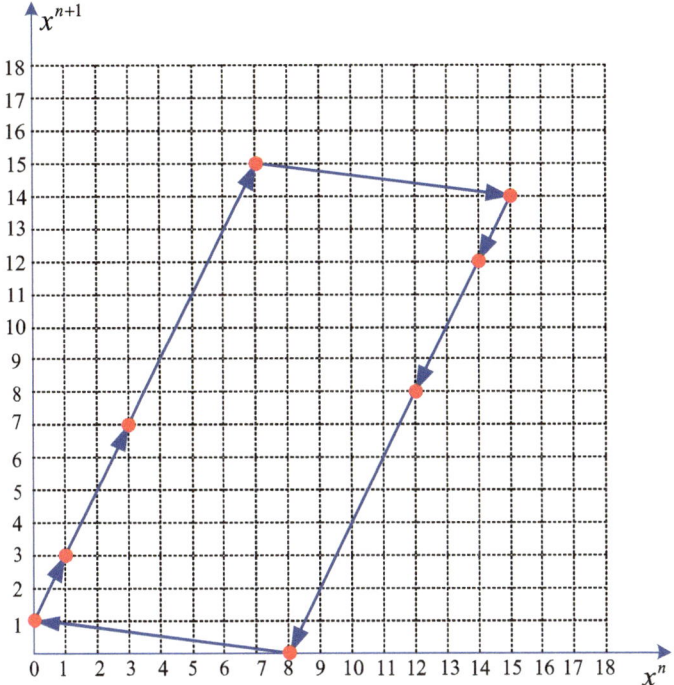

Fig. 2.3 The nonlinearization between the input and the output by $x_0^n \to x_1^{n+1} \to x_2^{n+2} \to x_3^{n+3}$

$$
\begin{cases}
x^0 = (x_3^0 x_2^0 x_1^0 x_0^0)_2 = (0000)_2 = 0; \\
x^1 = (x_3^1 x_2^1 x_1^1 x_0^1)_2 = (0001)_2 = 1; \quad x^2 = (x_3^2 x_2^2 x_1^2 x_0^2)_2 = (0011)_2 = 3; \\
x^3 = (x_3^3 x_2^3 x_1^3 x_0^3)_2 = (0111)_2 = 7; \quad x^4 = (x_3^4 x_2^4 x_1^4 x_0^4)_2 = (1111)_2 = 15; \\
x^5 = (x_3^5 x_2^5 x_1^5 x_0^5)_2 = (1110)_2 = 14; \quad x^6 = (x_3^6 x_2^6 x_1^6 x_0^6)_2 = (1100)_2 = 12; \\
x^7 = (x_3^7 x_2^7 x_1^7 x_0^7)_2 = (1000)_2 = 8; \quad x^8 = (x_3^8 x_2^8 x_1^8 x_0^8)_2 = (0000)_2 = 0;
\end{cases}
$$

$$(2.16)$$

where $()_2$ means that the value is in binary form.

Using the above Eq. (2.16), the nonlinear relationship between the input and the output of the system is shown in Fig. 2.3. And if the Not operation for x is run in other order, the relationship between the input and the output must also be nonlinear. For example, the nonlinear relationship between the input and output of the same system in this loop $x_0^n \to x_2^{n+1} \to x_1^{n+2} \to x_3^{n+3}$ is shown in Fig. 2.4. Normally, if the Not operation for x is run in any random order, the relationship between the input and output of the chaotic system is still nonlinear.

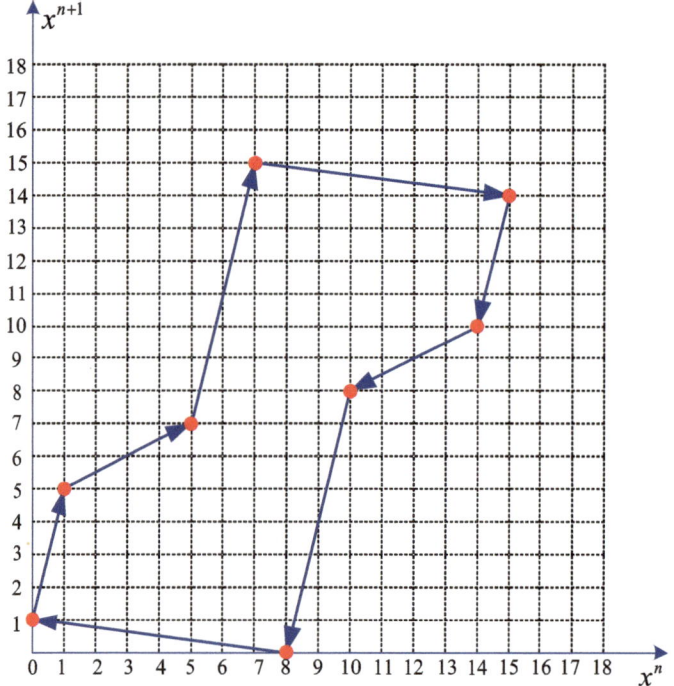

Fig. 2.4 The nonlinearization between the input and the output by $x_0^n \rightarrow x_2^{n+1} \rightarrow x_1^{n+2} \rightarrow x_3^{n+3}$

2.3   Network Analysis of the State Space of IDCS

In this section, the iterative function f is the vectorial negation; IDCSs with $N = 3$ and $N = 4$ are used to illustrate the property on connectivity.

2.3.1   The Corresponding State Transition Diagram and Its Connectivity Analysis for IDCS with $N = 3$

The relationship between input x^n and output x^{n+1} for IDCS with $N = 3$ is shown in Fig. 2.5, where its state transition is analyzed as follows.

1. When the input is $x^n = (000)_2 = 0$, the corresponding three possible outputs x^{n+1} are

$$x^{n+1} \in \left\{ \begin{array}{c} (100)_2 \\ (010)_2 \\ (001)_2 \end{array} \right\} \quad \Rightarrow x^{n+1} \in \left\{ \begin{array}{c} 4 \\ 2 \\ 1 \end{array} \right\} \tag{2.17}$$

as shown in the first column in Fig. 2.5.

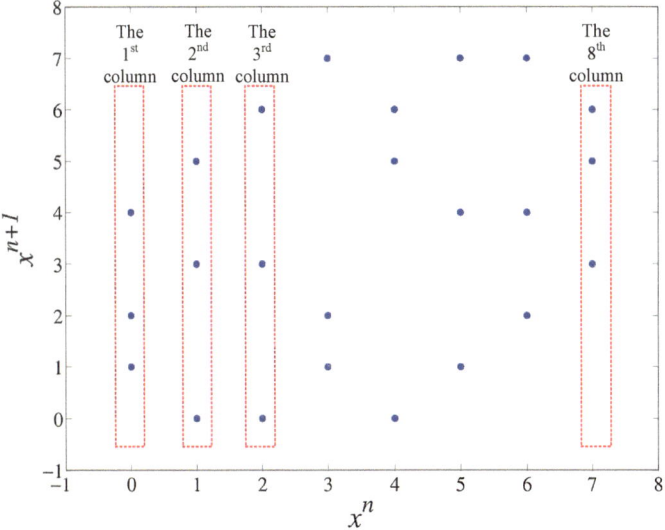

Fig. 2.5 The relationship between input x^n and output x^{n+1} for IDCS with $N = 3$

2. When the input is $x^n = (001)_2 = 1$, the corresponding three possible outputs x^{n+1} are

$$x^{n+1} \in \left\{ \begin{array}{c} (101)_2 \\ (011)_2 \\ (000)_2 \end{array} \right\} \Rightarrow x^{n+1} \in \left\{ \begin{array}{c} 5 \\ 3 \\ 0 \end{array} \right\} \tag{2.18}$$

as shown in the second column in Fig. 2.5.
3. When the input is $x^n = (010)_2 = 2$, the corresponding three possible outputs x^{n+1} are

$$x^{n+1} \in \left\{ \begin{array}{c} (110)_2 \\ (000)_2 \\ (011)_2 \end{array} \right\} \Rightarrow x^{n+1} \in \left\{ \begin{array}{c} 6 \\ 0 \\ 3 \end{array} \right\} \tag{2.19}$$

as shown in the third column in Fig. 2.5.
4. The remaining values $x^n = (011)_2 = 3$, $x^n = (100)_2 = 4$, $x^n = (101)_2 = 5$, and $x^n = (110)_2 = 6$, may be deduced by analogy.
5. Finally, when the input is $x^n = (111)_2 = 7$, the corresponding three possible outputs x^{n+1} are

$$x^{n+1} \in \left\{ \begin{array}{c} (011)_2 \\ (101)_2 \\ (110)_2 \end{array} \right\} \Rightarrow x^{n+1} \in \left\{ \begin{array}{c} 3 \\ 5 \\ 6 \end{array} \right\} \tag{2.20}$$

as shown in the eighth column in Fig. 2.5.

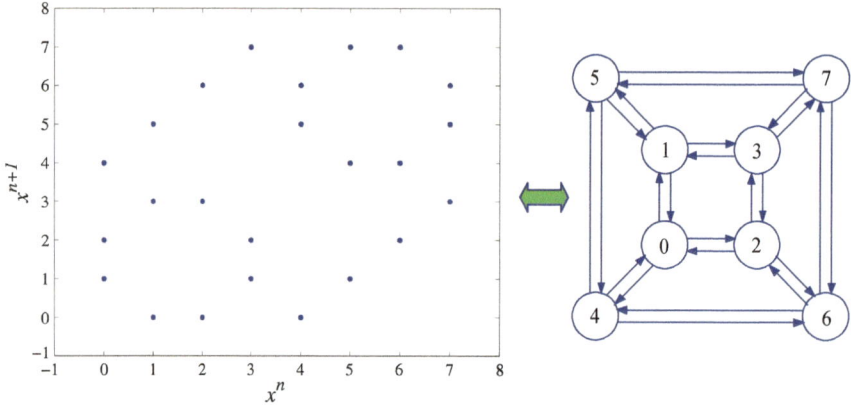

Fig. 2.6 The strongly connected graph for IDCS with $N = 3$

If the eight states 0, 1, 2, 3, 4, 5, 6, and 7 are marked as vertices ⓪, ①, ②, ③, ④, ⑤, ⑥, and ⑦, and the vertex and its interval are mapped to another one, then the state network of IDCS is built up as shown in Fig. 2.6.

2.3.2 The Corresponding State Transition Diagram and Its Connectivity Analysis for IDCS with $N = 4$

The relationship between input x^n and output x^{n+1} for IDCS with $N = 4$ is shown in Fig. 2.7, where its state transition is analyzed as

1. When the input is $x^n = (0000)_2 = 0$, the corresponding four possible outputs x^{n+1} are

$$x^{n+1} \in \left\{ \begin{array}{c} (1000)_2 \\ (0100)_2 \\ (0010)_2 \\ (0001)_2 \end{array} \right\} \Rightarrow x^{n+1} \in \left\{ \begin{array}{c} 8 \\ 4 \\ 2 \\ 1 \end{array} \right\} \tag{2.21}$$

   as shown in the first column in Fig. 2.7.

2. When the input is $x^n = (0001)_2 = 1$, the corresponding four possible outputs x^{n+1} are

$$x^{n+1} \in \left\{ \begin{array}{c} (1001)_2 \\ (0101)_2 \\ (0011)_2 \\ (0000)_2 \end{array} \right\} \Rightarrow x^{n+1} \in \left\{ \begin{array}{c} 9 \\ 5 \\ 3 \\ 0 \end{array} \right\} \tag{2.22}$$

   as shown in the second column of Fig. 2.7.

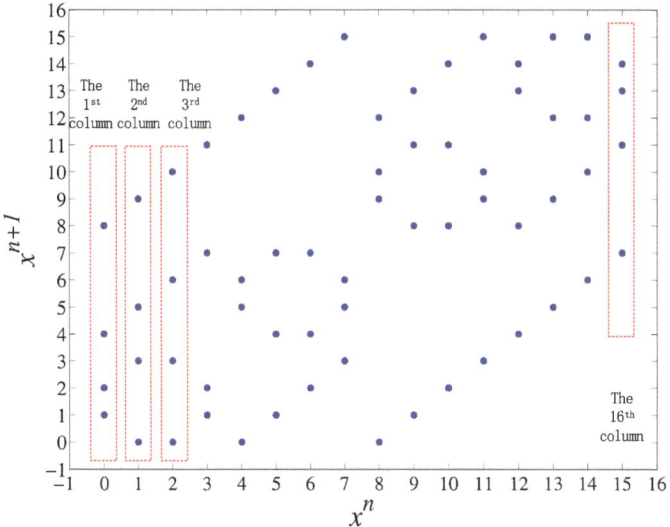

Fig. 2.7 The relationship between input x^n and output x^{n+1} for IDCS with $N = 4$

3. When the input is $x^n = (0010)_2 = 2$, the corresponding four possible outputs x^{n+1} are

$$x^{n+1} \in \begin{Bmatrix} (1010)_2 \\ (0110)_2 \\ (0000)_2 \\ (0011)_2 \end{Bmatrix} \Rightarrow x^{n+1} \in \begin{Bmatrix} 10 \\ 6 \\ 0 \\ 3 \end{Bmatrix} \tag{2.23}$$

as shown in the third column in Fig. 2.7.

4. The case of $x^n = (0011)_2 = 3$, $x^n = (0100)_2 = 4$, ..., $x^n = (1110)_2 = 14$, may be deduced by analogy.

5. When the input is $x^n = (1111)_2 = 15$, the corresponding four possible outputs x^{n+1} are

$$x^{n+1} \in \begin{Bmatrix} (0111)_2 \\ (1011)_2 \\ (1101)_2 \\ (1110)_2 \end{Bmatrix} \Rightarrow x^{n+1} \in \begin{Bmatrix} 7 \\ 11 \\ 13 \\ 14 \end{Bmatrix} \tag{2.24}$$

as shown in the sixteenth column in Fig. 2.7.

If the states 0, ..., 14, and 15 are marked as vertex ⓪, ..., ⑮, and the vertex and its interval is mapped to another one, then the state network of IDCS is built up as shown in Fig. 2.8.

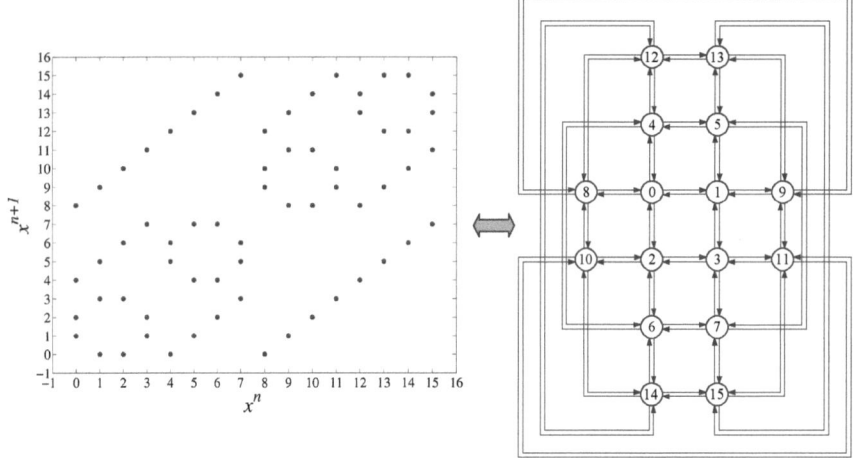

Fig. 2.8 The strongly connected graph for IDCS with $N = 4$

2.4 Circuit Implementation of IDCS

In this section, an IDCS circuit is designed, which consists of several submodules: uniform noise signal generator, noise voltage converter, sample-hold circuit, decoder circuit, iterative function circuit, and digital-to-analog converter. Finally, both the validity and practicability are verified by the experimental results.

The uniform noise signal generator is provided in Fig. 2.9. It uses an MM5837 broadband white-noise generator with a 3 dB per octave filter from 10 Hz to 40 kHz to give noise output $\xi(t)$, which has a flat spectral distribution over the entire audio band from 20 Hz to 20 kHz. Output is about $1 V_{P-P}$ of noise riding on the 8.5 V level. The parameters of components in Fig. 2.9 are capacitances $C_1 = 100 \mu F$, $C_2 = 1 \mu F$, $C_3 = 0.27 \mu F$, $C_4 = C_5 = 0.047 \mu F$, $C_6 = 0.033 \mu F$, and resistances $R_1 = 6.8 k\Omega$, $R_2 = 3 k\Omega$, $R_3 = 1 k\Omega$, and $R_4 = 300 \Omega$. In Fig. 2.9, the uniform

Fig. 2.9 The uniform noise signal generator. (©World Scientific Publishing Co Pte Ltd 2014, Reproduced with permission from [1])

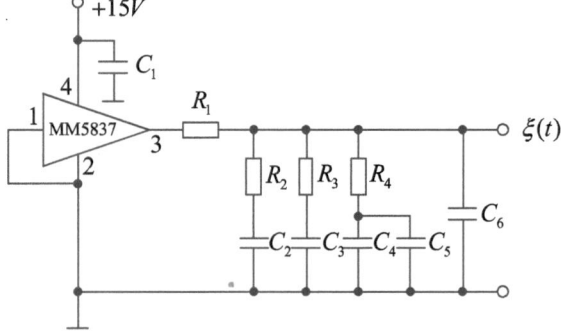

Fig. 2.10 The noise voltage converter. (©World Scientific Publishing Co Pte Ltd 2014, Reproduced with permission from [1])

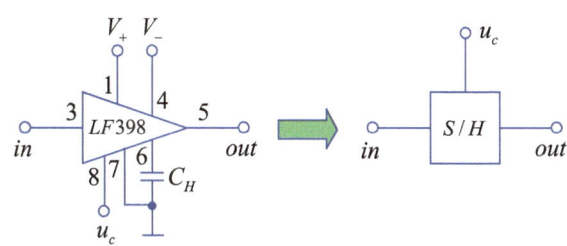

Fig. 2.11 The sample-hold circuit. (©World Scientific Publishing Co Pte Ltd 2014, Reproduced with permission from [1])

noise signal generator output is about $1 V_{P-P}$ of noise riding on the 8.5 V level, therefore it should be converted to a $0 \sim 4$ V uniform noise signal. A noise voltage converter is shown in Fig. 2.10. The values of each resistance in Fig. 2.10 are $R_5 = R_6 = R_8 = R_9 = 10 \, \mathrm{k\Omega}$, $R_7 = 40 \, \mathrm{k\Omega}$. The noise output $\xi(t) = 0 \sim 4$ V. A sample-hold circuit is shown in Fig. 2.11, in which the chip model is LF398. The supply voltage is $V_+ = +15$ V, $V_- = -15$ V. In Fig. 2.11, the 3-pin is for analog signal input, the 5-pin is an output; capacitor $C_F = 0.01 \sim 0.1 \, \mu\mathrm{F}$ (0.022 μF is used here). u_c is a square wave signal with frequency $1 \sim 5 \mathrm{kHz}$ (4 kHz is used here), the amplitude of output is $-5 \sim 5$ V. Note that when C_F is enlarged, then the frequency of u_c is reduced, and thus the iterations are slower. Conversely, if C_F is smaller, then the frequency of u_c may be higher, and so the iteration speed is faster. Due to the speed of the device itself, the iteration speed has an upper limitation. When doing experiments, C_F should be a suitable value, the same for the frequency of u_c, that prevents work abnormality. The decoding circuit is shown in Fig. 2.12, and the corresponding comparator circuit is described in Fig. 2.13. The values for each resistance are $R_{10} = 13.5 \, \mathrm{k\Omega}$, $R_{11} = 1 \, \mathrm{k\Omega}$, $R_{12} = 10 \, \mathrm{k\Omega}$, $R_{13} = 40 \, \Omega$, and $R_{14} = R_{15} = R_{16} = 10 \, \mathrm{k\Omega}$, whereas the voltage for inverting the voltage shifter is $E = 4$. According to Fig. 2.13, the logical relationship for input and output of the comparator is

$$\begin{cases} \text{if } \eta(n) > U_i, \text{ then } \eta_i = 1 \ (4\,\mathrm{V}), \\ \text{if } \eta(n) < U_i, \text{ then } \eta_i = 0 \ (0\,\mathrm{V}). \end{cases} \tag{2.25}$$

According to Fig. 2.12, the input-output relationship of the decoding circuit is:

1. When $3\mathrm{V} < \eta(t) \leqslant 4\mathrm{V}$, then $\eta_3 = \eta_2 = \eta_1 = \eta_0 = 1$, therefore

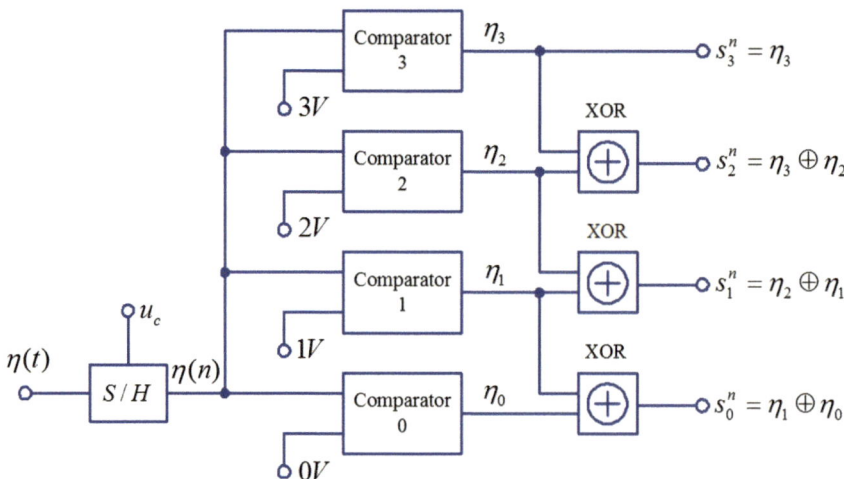

Fig. 2.12 The decoder circuit. (©World Scientific Publishing Co Pte Ltd 2014, Reproduced with permission from [1])

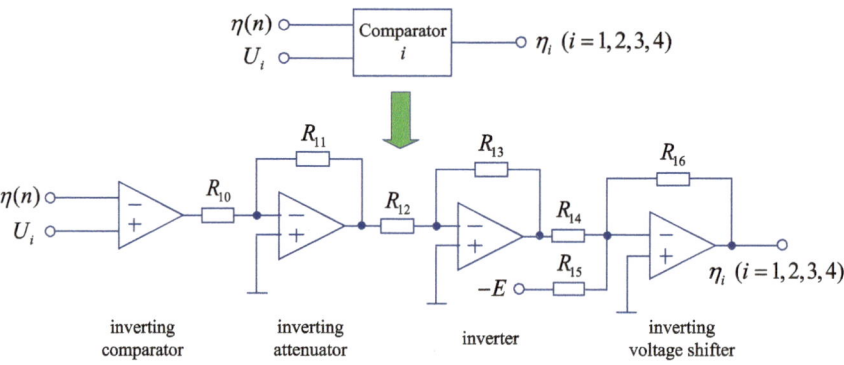

Fig. 2.13 The comparator. (©World Scientific Publishing Co Pte Ltd 2014, Reproduced with permission from [1])

$$\begin{cases} s_3^n = \eta_3 = 1, \\ s_2^n = \eta_3 \oplus \eta_2 = 1 \oplus 1 = 0, \\ s_1^n = \eta_2 \oplus \eta_1 = 1 \oplus 1 = 0, \\ s_0^n = \eta_1 \oplus \eta_0 = 1 \oplus 1 = 0. \end{cases} \tag{2.26}$$

2. When $2V < \eta(t) \leqslant 3V$, then $\eta_3 = 0, \eta_2 = \eta_1 = \eta_0 = 1$, therefore

$$\begin{cases} s_3^n = \eta_3 = 0, \\ s_2^n = \eta_3 \oplus \eta_2 = 0 \oplus 1 = 1, \\ s_1^n = \eta_2 \oplus \eta_1 = 1 \oplus 1 = 0, \\ s_0^n = \eta_1 \oplus \eta_0 = 1 \oplus 1 = 0. \end{cases} \tag{2.27}$$

3. When $1V < \eta(t) \leqslant 2V$, then $\eta_3 = \eta_2 = 0$, $\eta_1 = \eta_0 = 1$, therefore

$$\begin{cases} s_3^n = \eta_3 = 0, \\ s_2^n = \eta_3 \oplus \eta_2 = 0 \oplus 0 = 0, \\ s_1^n = \eta_2 \oplus \eta_1 = 0 \oplus 1 = 1, \\ s_0^n = \eta_1 \oplus \eta_0 = 1 \oplus 1 = 0. \end{cases} \tag{2.28}$$

4. When $0V < \eta(t) \leqslant 1V$, then $\eta_3 = \eta_2 = \eta_1 = 0$, $\eta_0 = 1$, thus

$$\begin{cases} s_3^n = \eta_3 = 0, \\ s_2^n = \eta_3 \oplus \eta_2 = 0 \oplus 0 = 0, \\ s_1^n = \eta_2 \oplus \eta_1 = 0 \oplus 0 = 0, \\ s_0^n = \eta_1 \oplus \eta_0 = 0 \oplus 1 = 1. \end{cases} \tag{2.29}$$

It can be seen in Fig. 2.10 that the noise output satisfies $0V < \eta(t) \leqslant 4V$, and $\eta(t)$ is the random signal that follows an equal probability distribution (*i.e.*, uniform distribution) within the range $[0V, 4V]$. In other words, the values in these four intervals $([0V, 1V], [1V, 2V], [2V, 3V], [3V, 4V])$ are uniformly distributed, and the correspondence relationship between the size of s^n and the four intervals is:

$$\begin{cases} \text{if } \eta(t) \in [0V, 1V], \text{ then } s^n = 0, \\ \text{if } \eta(t) \in [1V, 2V], \text{ then } s^n = 1, \\ \text{if } \eta(t) \in [2V, 3V], \text{ then } s^n = 2, \\ \text{if } \eta(t) \in [3V, 4V], \text{ then } s^n = 3. \end{cases} \tag{2.30}$$

Through the above comparison, it is known that both s^n and $s_3^n s_2^n s_1^n s_0^n$ follow the uniform distribution, and the relationship between them satisfies:

$$\begin{cases} \text{if } \eta(t) \in [0V, 1V], \text{ then } s^n = 0 \Leftrightarrow s_3^n s_2^n s_1^n s_0^n = 0001, \\ \text{if } \eta(t) \in [1V, 2V], \text{ then } s^n = 1 \Leftrightarrow s_3^n s_2^n s_1^n s_0^n = 0010, \\ \text{if } \eta(t) \in [2V, 3V], \text{ then } s^n = 2 \Leftrightarrow s_3^n s_2^n s_1^n s_0^n = 0100, \\ \text{if } \eta(t) \in [3V, 4V], \text{ then } s^n = 3 \Leftrightarrow s_3^n s_2^n s_1^n s_0^n = 1000. \end{cases} \tag{2.31}$$

Set $N = 4$, and get the basic iterative function for IDCS,

$$x_i^n = \begin{cases} x_i^{n-1} & \text{if } i \neq s^n, \\ (f(x^{n-1}))_i = \overline{x_i^{n-1}} & \text{if } i = s^n. \end{cases} \tag{2.32}$$

where $s^n \in \{0, 1, 2, \ldots N - 1\} = 0, 1, 2, 3$ and $i = 0, 1, 2, 3$. By comparing Eqs. (2.26)–(2.32), Eq. (2.32) is equivalent to another type of mathematical expression as

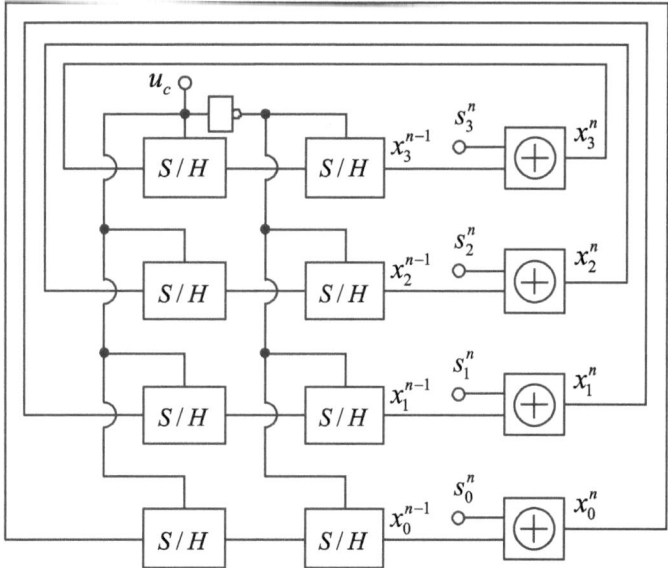

Fig. 2.14 The iterative function circuit. (©World Scientific Publishing Co Pte Ltd 2014, Reproduced with permission from [1])

$$\begin{cases} x_3^n = x_3^{n-1} \oplus s_3^n, \\ x_2^n = x_2^{n-1} \oplus s_2^n, \\ x_1^n = x_1^{n-1} \oplus s_1^n, \\ x_0^n = x_0^{n-1} \oplus s_0^n. \end{cases} \quad (2.33)$$

where

$$\begin{cases} s^n = 0 \Leftrightarrow s_3^n s_2^n s_1^n s_0^n = 0001, \\ s^n = 1 \Leftrightarrow s_3^n s_2^n s_1^n s_0^n = 0010, \\ s^n = 2 \Leftrightarrow s_3^n s_2^n s_1^n s_0^n = 0100, \\ s^n = 3 \Leftrightarrow s_3^n s_2^n s_1^n s_0^n = 1000. \end{cases} \quad (2.34)$$

Equation (2.33) totally corresponds to the chaotic iterations that have been studied in the first part of this chapter. According to this Eq. (2.33), we obtain the corresponding circuit design iteration equation shown in Fig. 2.14. The digital-to-analog converter is shown in Fig. 2.15, where $R_{17} = 10\,\text{k}\Omega$, $R_{18} = 2\,\text{k}\Omega$, $R_{19} = 60\,\text{k}\Omega$, and $R_{20} = R_{21} = 10\,\text{k}\Omega$. A DAC0832 is used; it should be configured to allow the analog output x^n continuously to reflect the state of an applied digital $D_3 D_2 D_1 D_0 = x_3^n x_2^n x_1^n x_0^n$ on flow-through operation. The logic relationship is: when the input is $x_3^n x_2^n x_1^n x_0^n = 0000$, the output is $x^n = 0\,\text{V}$; when the input is $x_3^n x_2^n x_1^n x_0^n = 0001$, the output is $x^n = 1\,\text{V}; ...$; when the input is $x_3^n x_2^n x_1^n x_0^n = 1111$, the output is $x^n = 15\,\text{V}$. The above correspondence can be adjusted by the resistance R_{19}. Based on Figs. 2.9, 2.10, 2.11,

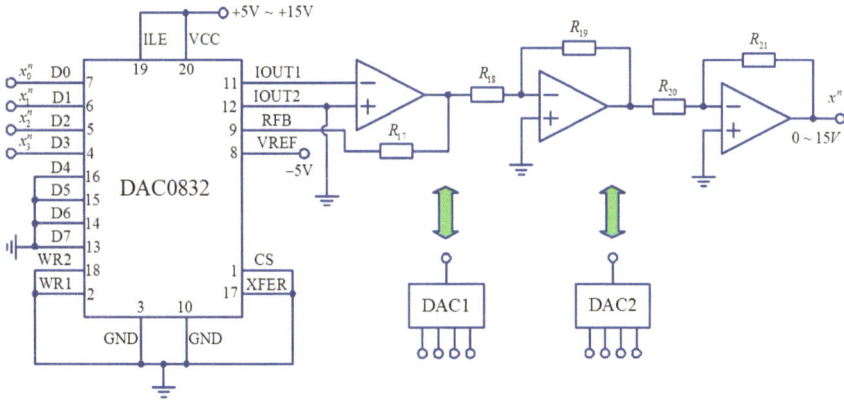

Fig. 2.15 The digital-to-analog converter. (©World Scientific Publishing Co Pte Ltd 2014, Reproduced with permission from [1])

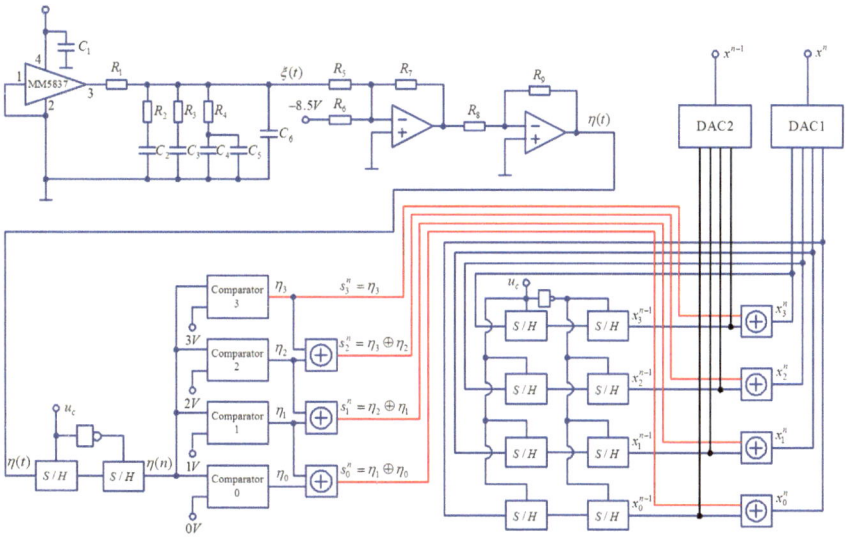

Fig. 2.16 The general IDCS circuit. (©World Scientific Publishing Co Pte Ltd 2014, Reproduced with permission from [1])

2.12, 2.13, 2.14 and 2.15, the whole basic IDCS circuit design is shown in Fig. 2.16, with experimental observations of IDCS as shown in Fig. 2.17, respectively. Finally, the hardware circuit implementation is provided in Fig. 2.18.

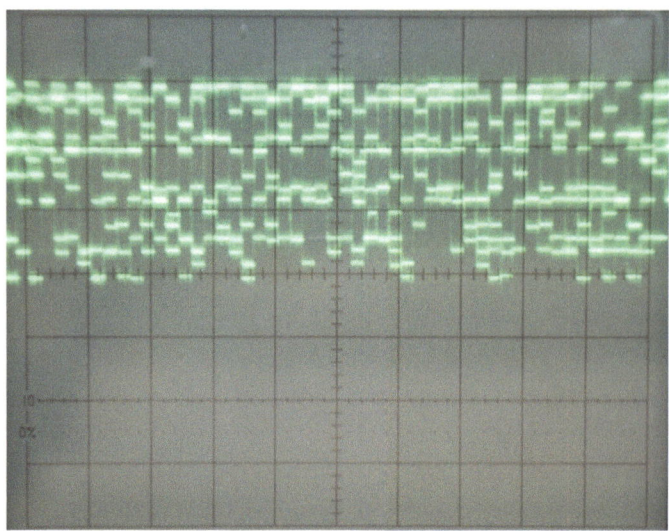

Fig. 2.17 The experimental observations of IDCS. (©World Scientific Publishing Co Pte Ltd 2014, Reproduced with permission from [1])

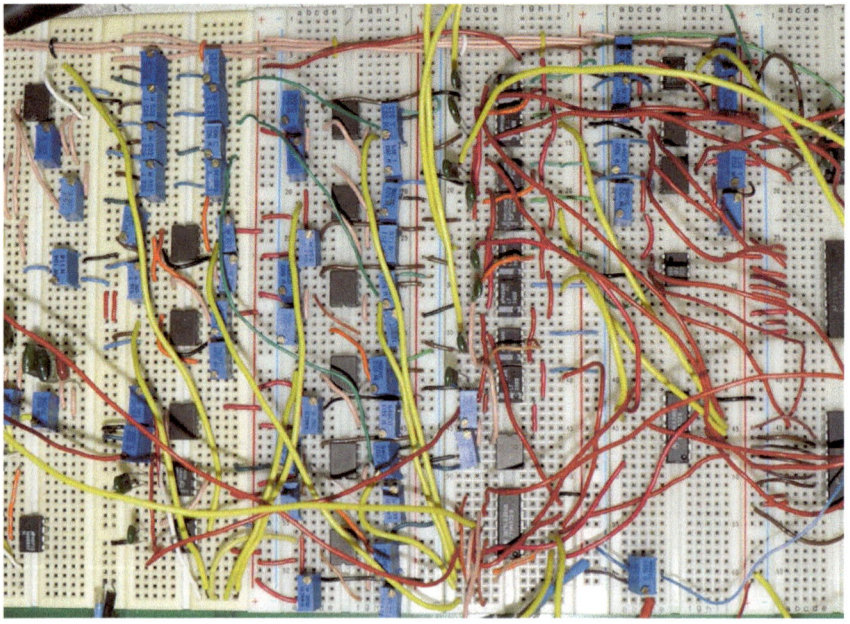

Fig. 2.18 The hardware circuit implementation of IDCS

References

1. Q. Wang, S. Yu, C. Guyeux, J. Bahi, X. Fang, Theoretical design and circuit implementation of integer domain chaotic systems. Int. J. Bifurc. Chaos **24**(10) (2014). p. Art. no. 1450128
2. Q. Wang, S. Yu, C. Guyeux, J. Bahi, X. Fang, Study on a new chaotic bitwise dynamical system and its FPGA implementation. Chin. Phys. B **24**(6) (2015). Art no. 60503
3. Q. Wang, S. Yu, C. Li, J. Lü, X. Fang, C. Guyeux, J. Bahi, Theoretical design and FPGA-based implementation of higher-dimensional digital chaotic systems. IEEE Trans. Circuits Syst. I **63**(3), 401–412 (2016)
4. J. Bahi, C. Guyeux, Q.X. Wang, A pseudo random numbers generator based on chaotic iterations. Application to watermarking, in *IEEE International Conference on Web Information Systems and Mining*, Sanya, China, October 2010, LNCS (2010), vol. 6318, pp. 202–211

Chapter 3
Chaotic Bitwise Dynamical Systems (CBDS)

3.1 Improvements of Chaotic Bitwise Dynamical Systems (CBDS)

In this section, we first recall the basic concept of real domain chaotic systems (RDCS) and integer domain chaotic systems (IDCS).

Let $N \in \{1, 2, \ldots\}$ be a positive integer, $\mathbb{B} = \{0, 1\}$ denote the set of Boolean numbers with its usual algebraic structure, and $\mathbb{B}^N$ be the set of binary vectors of size N.

Definition 3.1 Let $f : \mathbb{R} \to \mathbb{R}$ be a function, and x^n and x^{n-1} be two real numbers, which are represented as an infinite number of bits in their binary decomposition:

$$\begin{cases} x^{n-1} = x_{i_1} x_{i_2} \ldots x_{i_M} . x_{j_1} x_{j_2} \ldots x_{j_N} \ldots , \\ x^n = x_{k_1} x_{k_2} \ldots x_{k_L} . x_{l_1} x_{l_2} \ldots x_{l_P} \ldots , \end{cases} \tag{3.1}$$

whereas $x_{i_1} x_{i_2} \ldots x_{i_M}$ and $x_{j_1} x_{j_2} \ldots x_{j_N} \ldots$ are, respectively, the integral and fractional parts for x_{n-1}. Similarly, $x_{k_1} x_{k_2} \ldots x_{k_L}$ and $x_{l_1} x_{l_2} \ldots x_{l_P} \ldots$ are, respectively, the integral and fractional parts for x_n. The so-called iterative equation in RDCS is defined by:

$$x^{n-1} = x_{i_1} x_{i_2} \ldots x_{i_M} . x_{j_1} x_{j_2} \ldots x_{j_N} \ldots \to x^n = f(x^{n-1}) =$$
$$f(x^{n-1})_{k_1} f(x^{n-1})_{k_2} \ldots f(x^{n-1})_{k_L} . f(x^{n-1})_{l_1} f(x^{n-1})_{l_2} \ldots f(x^{n-1})_{l_P} \ldots , \tag{3.2}$$

where $f(x^{n-1})_k$ and $f(x^{n-1})_l$ are the kth and lth components of $f(x^{n-1})$ for the integral and fractional parts, respectively. In other words, at each iteration all the binary components of these real numbers are iterated.

Parts of this chapter were reproduced with permission from [1] ©World Scientific Publishing Co Pte Ltd 2014, [2] ©Chinese Physical Society and IOP Publishing Ltd 2015, and [3] ©IEEE 2016.

Q. Wang et al., *Design of Digital Chaotic Systems Updated by Random Iterations*, SpringerBriefs in Nonlinear Circuits, https://doi.org/10.1007/978-3-319-73549-8_3

Consider now a system with a finite number of N elements (or cells) $x = (x_{N-1} x_{N-2} \ldots x_0)$, so that $x_{N-1}, x_{N-2}, \ldots, x_0 \in \mathbb{B}$. Having N Boolean values for these cells leads to the definition of a particular state of the system $x = (x_{N-1}, x_{N-2}, \ldots, x_0) \in \mathbb{B}^N$. A one-sided infinite sequence of integers bounded by $N - 1$ is called a strategy: $\forall n \in \mathbb{N},\ s^n \in \{0, 1, 2, \ldots N - 1\}$.

Definition 3.2 ([1]) Let $f : \mathbb{B}^N \rightarrow \mathbb{B}^N$ be a function, $x^0 = (x_{N-1}^0, x_{N-2}^0, \ldots, x_0^0) \in \mathbb{B}^N$ be the initial condition, and $x^{n-1} = (x_{N-1}^{n-1}, x_{N-2}^{n-1}, \ldots, x_0^{n-1}) \in \mathbb{B}^N$ and $x^n = (x_{N-1}^n, x_{N-2}^n, \ldots, x_0^n) \in \mathbb{B}^N$ denote the $(n-1)$th and nth iteration, respectively. The so-called iterative equation in IDCSs is defined by:

$$x_i^n = \begin{cases} x_i^{n-1} & if\ i \neq s^n \\ (f(x^{n-1}))_i & if\ i = s^n \end{cases} \rightarrow$$
$$x^n = (x_{N-1}^{n-1}, x_{N-2}^{n-1}, \ldots, x_{s^n+1}^{n-1}, f(x^{n-1})_{s^n}, x_{s^n-1}^{n-1}, \ldots, x_0^{n-1}), \tag{3.3}$$

where $i \in 0, 1, 2, \ldots N - 1$, $n = 1, 2, \ldots$, and $f(x^{n-1})_{s^n}$ is the s^nth component of $f(x^{n-1})$.

In other words, at the nth iteration, only the s^nth element is updated by f.

The major difference between CBDS and the basic IDCS is that w^n is a subset ($w^n \subset \{0, 2, \ldots N - 1\}$) instead of a single integer ($s^n \in \{0, 2, \ldots N - 1\}$).

Definition 3.3 In the so-called CBDS, the iterative equation is defined as

$$x_i^n = \begin{cases} x_i^{n-1} & if\ i \notin w^n \\ (f(x^{n-1}))_i & if\ i \in w^n \end{cases} \rightarrow$$
$$x^n = (x_{N-1}^{n-1}, \ldots, x_{w_1^n+1}^{n-1}, f(x^{n-1})_{w_1^n}, x_{w_1^n-1}^{n-1}, \ldots, f(x^{n-1})_{w_2^n}, \ldots, f(x^{n-1})_{w_{|w^n|}^n}, \ldots, x_0^{n-1}), \tag{3.4}$$

where $i \in 0, 1, 2, \ldots N - 1$, $n = 1, 2, \ldots$, and w^n is a set, in which the value is bounded by $N - 1$ as $w^n = w_1^n, w_2^n, \ldots, w_{|w^n|}^n$, that is, $w_j^i \in \{0, 1, 2, \ldots N - 1\}$, where $j \in \{1, 2, \ldots |w^n|\}$. Let us remark that $|.|$ means the size of the set.

In other words, at the nth iteration, only the elements whose id is contained in the set w^n are updated [4].

The set w^n can be expressed as an integer s^n having N bits: the kth digit in the binary decomposition of s^n is 1 if and only if k belongs to w^n (the other digits are set to 0). That is, the size of the set $|w^n|$ is the number of 1s in the binary decomposition of s^n, and the elements in the set w^n are the positions of these 1s. Therefore, the iterative equation is:

$$x^n = (x^{n-1} \cdot \overline{s^n}) + (f(x^{n-1}) \cdot s^n), \tag{3.5}$$

where the operators " $\cdot$ ", "$\overline{(\cdot)}$", and " $+$ " denote bitwise AND, bitwise NOT (negation), and bitwise OR, respectively. This formula means that the kth bit of state x^n (in binary form) is updated by $f(x^{n-1})_k$ if and only if the kth digit in the binary decomposition of s^n is 1. Remark that in our previous research work the iterative function f has often been the vectorial negation [1], given by:

$$f(x^{n-1}) = \overline{x^{n-1}}. \tag{3.6}$$

With such a choice, a more specific formula can be obtained:

$$F_f(s, x) = (x^{n-1} \cdot \overline{s^n}) + (\overline{x^{n-1}} \cdot s^n) = x \oplus s, \tag{3.7}$$

where $\oplus$ denotes the bitwise XOR.

Let $E = (s, x)$ be a couple constituted by a chaotic strategy and a binary digit; consider the phase space:

$$\mathcal{E} = \{0, 1, 2 \ldots 2^N - 1\}^\infty \times \mathbb{B}^N \tag{3.8}$$

and the map defined on $\mathcal{E}$ with $E = (s, x) \in \mathcal{E}$:

$$G_f(E) = G_f((s, x)) = (\sigma(s), F_f(i(s), x)), \tag{3.9}$$

where $i(s) = s^1$ and $\sigma^k(s) = \underbrace{\sigma \circ \sigma \circ \ldots \circ \sigma}_{k}(s), k = 1, 2 \ldots$ is the result of shifting k integers in the one-sided infinite sequence $s = (s^1 s^2 \ldots s^n \ldots)$ to the left:

$$\sigma^k(s) = s^{k+1} s^{k+2} \ldots s^n \ldots, \quad (k = 1, 2 \ldots). \tag{3.10}$$

Then the iterative equation proposed in Definition 3.3 can be rewritten as

$$E^0 \in \mathcal{E} \ and \ \forall k \in \mathbb{N}, \ E^{k+1} = G_f(E^k). \tag{3.11}$$

To study Devaney's chaos property, a distance between two points (s, x) and $(\hat{s}, \hat{x})$ of $\mathcal{E}$ must be defined. Let us introduce:

$$d((s, x), (\hat{s}, \hat{x})) = d_s(s, \hat{s}) + d_x(x, \hat{x}), \tag{3.12}$$

where $s = s^1 s^2 s^3 \ldots s^n \ldots$ and $\hat{s} = \hat{s}^1 \hat{s}^2 \hat{s}^3 \ldots \hat{s}^n \ldots$ are one-sided infinite sequences of integers, and x and $\hat{x}$ are binary digits of N bits. More precisely, the distance between s and $\hat{s}$ is:

$$d_s(s, \hat{s}) = \sum_{k=1}^{\infty} \frac{s^k \oplus \hat{s}^k}{2^{Nk}} \in [0, 1], \tag{3.13}$$

where $s^k \oplus \hat{s}^k$ is for the symmetric difference. Finally, following the 1-norm distance, the distance between x and $\hat{x}$ is:

$$d_x(x, \hat{x}) = x^k \oplus \hat{x}^k. \tag{3.14}$$

Remark that d is defined as the sum of two other distances. Before investigating the chaotic properties of CBDS, the following lemma is introduced.

Lemma 3.1 *Let* $s = s^1 s^2 s^3 \ldots s^n \ldots$ *and* $\hat{s} = \hat{s}^1 \hat{s}^2 \hat{s}^3 \ldots \hat{s}^n \ldots,$ *where* $s^k, \hat{s}^k$ $\in \{0, 1, 2, \ldots, 2^N - 1\}$ *for* $k = 1, 2, \ldots$ *If* $s^i = \hat{s}^i$ *for* $i = 1, 2, \ldots, n,$ *then* $d_s(s, \hat{s}) \le$ $\frac{1}{2^{Nn}}.$ *Conversely, if* $d_s(s, \hat{s}) \le \frac{1}{2^{Nn}},$ *then* $s^i = \hat{s}^i$ *for* $i = 1, 2, \ldots, n.$

Proof If $s^i = \hat{s}^i$ for $i = 1, 2, \ldots, n,$ then

$$
\begin{aligned}
d_s(s, \hat{s}) &= \sum_{i=1}^{n} \frac{s^i \oplus \hat{s}^i}{2^{Ni}} + \sum_{i=n+1}^{\infty} \frac{s^i \oplus \hat{s}^i}{2^{Ni}} \\
&= \sum_{i=n+1}^{\infty} \frac{s^i \oplus \hat{s}^i}{2^{Ni}} \le \sum_{i=n+1}^{\infty} \frac{2^N - 1}{2^{Ni}} = (2^N - 1) \times \frac{\frac{1}{2^{Nn+N}}}{1 - \frac{1}{2^N}} \qquad (3.15) \\
&\le \frac{1}{2^{Nn}}.
\end{aligned}
$$

Conversely, and due to the definition of the proposed distance: for any $m \le n,$ if $s^m \ne \hat{s}^m,$ then $d(s, \hat{s}) \ge \frac{1}{2^{Nn}}.$ The contraposition is the desired result: if $d_s(s, \hat{s}) \le \frac{1}{2^{Nn}},$ then $s^1 = \hat{s}^1.$

3.2 Proof of Chaos for CBDS

In this section, the chaotic behavior of CBDS is proven according to Devaney's definition.

3.2.1 Dense Periodic Points

Theorem 3.1 *The periodic points of* G_f *are dense in* $(\mathscr{E}, d).$

Proof Let $(\hat{s}, \hat{x}) \in \mathscr{E},$ and $\varepsilon > 0.$ We are looking for a periodic point $(\tilde{s}, \tilde{x}) \in \mathscr{E}$ that satisfies $d((\tilde{s}, \tilde{x}), (\hat{s}, \hat{x})) < \varepsilon.$

1. The general form of $(\hat{s}, \hat{x})$ is

$$
(\hat{s}, \hat{x}) = ((\hat{s}^1 \hat{s}^2 \ldots \hat{s}^{k_0} \ldots \hat{s}^n \ldots), \hat{x}) \in \mathscr{E}. \qquad (3.16)
$$

As ε can be strictly lower than 1, we must choose $\tilde{x} = \hat{x},$ because if $\tilde{x} \ne \hat{x},$ and $d_x(\tilde{x}, \hat{x}) \ge 1,$ then $d((\hat{s}, \hat{x}), (\tilde{s}, \tilde{x})) > 1.$

2. Let us define $k_0 = \lfloor \dfrac{-log_2(\varepsilon)}{N} \rfloor + 1$ according to the previous Lemma 3.1, and let us consider that the first k_0 elements of $\hat{s}$ and $\tilde{s}$ are the same. $\forall \varepsilon < 1,$ an integer k_0 can always be found to satisfy the relation $d_s(\hat{s}, \tilde{s}) < 2^{-Nk_0} < \varepsilon.$

3. If after the k_0th iteration, we have

$$(G_f^{k_0}(\tilde{s}, \tilde{x}))_x = \tilde{x} = \hat{x}, \tag{3.17}$$

then a cycle point $(\tilde{s}, \tilde{x}) = ((\hat{s}^1 \hat{s}^2 \ldots \hat{s}^{k_0} \hat{s}^1 \hat{s}^2 \ldots \hat{s}^{k_0} \ldots), \hat{x}) \in \mathscr{E}$ is found that satisfies

$$(\tilde{s}, \tilde{x}) = G_f^{k_0}((\tilde{s}, \tilde{x})) \tag{3.18}$$

making

$$d((\tilde{s}, \tilde{x}), (\hat{s}, \hat{x})) = d_s(\tilde{s}, \hat{s}) + d_x(\tilde{x}, \hat{x}) = d_s(\tilde{s}, \hat{s}) < \varepsilon \tag{3.19}$$

true.

4. Suppose that, after the k_0th iteration, we have

$$G_f^{k_0}(\tilde{s}, \tilde{x}))_x \neq \tilde{x}. \tag{3.20}$$

To obtain that, after one more iteration, the following condition is met:

$$(G_f^{k_0+1}(\tilde{s}, \tilde{x}))_x = \tilde{x} = \hat{x},$$

we must set: $s^{k_0+1} = (G_f^{k_0}(\tilde{s}, \hat{x}))_x \oplus \hat{x}$. Then, within the range ε of the point $(\hat{s}, \hat{x})$, one can find the periodic point:

$$(\tilde{s}, \tilde{x}) = ((\hat{s}^1 \hat{s}^2 \ldots \hat{s}^{k_0} s^{k_0+1} \ldots \hat{s}^1 \hat{s}^2 \ldots \hat{s}^{k_0} s^{k_0+1} \ldots), \tilde{x}) = G_f^{k_0+1}(\tilde{s}, \tilde{x}) \tag{3.21}$$

making

$$d((\tilde{s}, \tilde{x}), (\hat{s}, \hat{x})) = d_s(\tilde{s}, \hat{s}) + d_x(\tilde{x}, \hat{x}) = d_s(\tilde{s}, \hat{s}) < \varepsilon \tag{3.22}$$

true.

In summary, the periodic points of G_f are dense in $\mathscr{E}$.

3.2.2 Transitive Property

Theorem 3.2 G_f is a transitive map in $(\mathscr{E}, d)$.

Proof Consider two nonempty open sets U_A and U_B, and $(s_A, x_A) \in U_A$, $(s_B, x_B) \in U_B$. U_A and U_B are open, and we take a place in a metric space, thus there exist real numbers $r_A > 0$ and $r_B > 0$ such that the open ball $\mathscr{B}_A$ with center (s_A, x_A) and radius r_A is inside U_A (resp., the open ball $\mathscr{B}_B$ with center (s_B, x_B) and radius r_B is inside U_B). Without loss of generality, we can suppose that $r_A < 1$.

1. We introduce the following notations.

$$(s_A, x_A) = ((s_A^1 s_A^2 \ldots s_A^{n_0} \ldots s_A^n \ldots), x_A) \in U_A \subseteq \mathscr{E}$$

and

$$(s_B, x_B) = ((s_B^1 s_B^2 \ldots s_B^n \ldots), x_B) \in U_B \subseteq \mathscr{E}.$$

2. Let $(\tilde{s}, \tilde{x}) \in U_A$. If $\tilde{x} \neq x_A$, then $d_x(\tilde{x}, x_A) \geqslant 1$, and thus $d((s_A, x_A), (\tilde{s}, \tilde{x})) > 1$. Consequently, if $(\tilde{s}, \tilde{x}) \in \mathscr{B}_A$, then $d((s_A, x_A), (\tilde{s}, \tilde{x})) < 1$, and thus $\tilde{x} = x_A$.
3. If we demand that the first k_0 elements of $\tilde{s}$ are equal to those from s_A, then we obtain $d_s(s_A, \tilde{s}) < 2^{-Nk_0}$. And for the given r_A, an integer k_0 (i.e., a sequence $\tilde{s}$) can always be found to achieve $d_s(s_A, \tilde{s}) < 2^{-Nk_0} < r_A$.
4. If after k_0 iterations, the following condition is satisfied,

$$(G_f^{k_0}(\tilde{s}, \tilde{x}))_x = x_B, \tag{3.23}$$

then $n_0 = k_0$ and $(\tilde{s}, \tilde{x}) = ((s_A^1 s_A^2 \ldots s_A^{n_0} s_B^1 s_B^2 \ldots s_B^n \ldots), x_A) \in U_A$ has been found that satisfies:

$$G_f^{n_0}(\tilde{s}, \tilde{x}) = (s_B, x_B) \in G_f^{n_0}(U_A) \cap U_B \tag{3.24}$$

making

$$G_f^{n_0}(U_A) \cap U_B \neq \varnothing \tag{3.25}$$

true.
5. If after the k_0th iteration, $(G_f^{k_0}(\tilde{s}, \tilde{x}))_x \neq x_B$, then now define

$$s^{k_0+1} = (G_f^{k_0}(\tilde{s}, \tilde{x}))_x \oplus x_B,$$

therefore the point $(\tilde{s}, \tilde{x}) = ((s_A^1 s_A^2 \ldots s_A^{k_0} s^{k_0+1} s_B^1 s_B^2 \ldots s_B^n \ldots), x_A) \in U_A$ satisfies $G_f^{n_0}(\tilde{s}, \tilde{x}) = (s_B, x_B) \in G_f^{n_0}(U_A) \cap U_B$ with $n_0 = k_0 + 1$, making the claim

$$G_f^{n_0}(U_A) \cap U_B \neq \varnothing \tag{3.26}$$

true.

In summary, $(\mathscr{E}, d)$ is transitive.

Because of dense periodic points and transitivity, and according to Definition 1.1 and Theorem 1.1 in Chap. 1, we can conclude that CBDS is chaotic in the sense of Devaney.

3.3   Uniformity

Definition 3.4 (*Uniform probability index*) Let γ be a random variable with a Bernoulli distribution. Therefore

$$Y = \frac{\min\{P(\gamma = 0), P(\gamma = 1)\}}{\max\{P(\gamma = 0), P(\gamma = 1)\}}, Y \in (0, 1] \tag{3.27}$$

is called the uniform probability index of γ.

The uniform probability index Y characterizes the uniformity of the binary sequence. In particular, if $Y = 1$, then γ is uniformly distributed. The larger Y is, the more uniform it is for γ.

Theorem 3.3 *Let α be a binary sequence having a Bernoulli distribution and β be a binary sequence defined by:* $\forall n, \beta^n = \beta^{n-1} \oplus \alpha^{n-1}$, *where $\alpha^n, \beta^n \in \{0, 1\}$. Then the uniformity of β is independent of α, in the sense that $Y_\beta = 1$.*

Proof Let $P(\beta = 0) = b$ and $P(\alpha = 0) = a$. According to $\beta^n = \beta^{n-1} \oplus \alpha^{n-1}$,

$$P(\beta = 0) = P(\alpha = 0, \beta = 0) + P(\alpha = 1, \beta = 1)$$
$$P(\beta = 0) = P(\alpha = 0)P(\beta = 0) + P(\alpha = 1)P(\beta = 1)$$
$$b = ab + (1 - a)(1 - b) \tag{3.28}$$
$$b(1 - a) = (1 - a)(1 - b)$$
$$b = 1/2.$$

It can also be proved the other way,

$$P(\beta = 1) = P(\alpha = 0, \beta = 1) + P(\alpha = 1, \beta = 0)$$
$$1 - b = a(1 - b) + (1 - a)b \tag{3.29}$$
$$(1 - b)(1 - a) = b(1 - a)$$
$$b = 1/2.$$

Through the above equations, we proved that $b = 1/2$ and $Y = 1$

Corollary 3.1 *Let s be a sequence of independent variables of integers having N bits and x be a sequence of random variables defined by $x^n = x^{n-1} \oplus s^n$. Then the uniformity of x is independent of s with $P(x = i) = 1/2^N$, where $i = 0, 1, 2, \ldots, 2^N - 1$.*

Proof This corollary is an obvious consequence of the aforementioned theorem. Take $P(x = 0)$ as an example:

$$P(x = 0) = P(x_{N-1}x_{N-2}\ldots x_0 = 00\ldots 0) = P(x_{N-1} = 0) \cdot P(x_{N-2} = 0) \cdot \ldots P(x_0 = 0).$$

For $j = 0, 1, \ldots, N - 1$,

$$P(x_j = 0) = P(s_j = 0)P(x_j = 0) + P(s_j = 1)P(x_j = 1).$$

According to $P(s_j = 0) + P(s_j = 1) = 1$ and $P(x_j = 0) + P(x_j = 1) = 1$, then

$$P(x_j = 0) = 1/2.$$

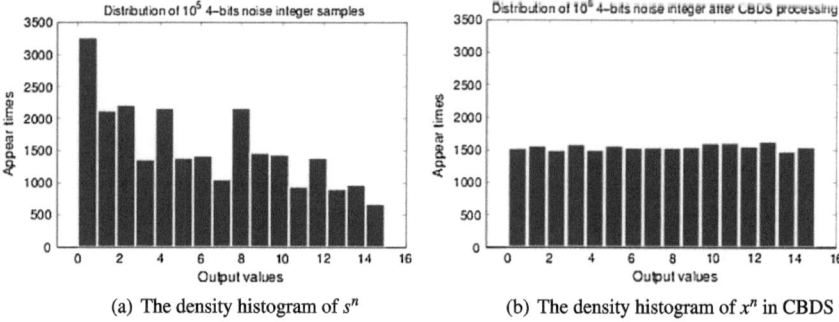

(a) The density histogram of s^n (b) The density histogram of x^n in CBDS

Fig. 3.1 The density histogram. **a** The density histogram of s^n. **b** The density histogram of x^n in CBDS. (©Chinese Physical Society and IOP Publishing Ltd 2015, Reproduced with permission from [2])

Thus,

$$P(x = 0) = P(x_{N-1} = 0) \cdot P(x_{N-2} = 0) \cdot \ldots P(x_0 = 0) = 1/2^N.$$

The rest $P(x = 1) = P(x = 2) = \ldots = P(x = 2^N - 1) = 1/2^N$ may be deduced by analogy.

In order to evaluate this conclusion, density histograms have been computed. In these experiments, the length of s^n and x^n is $N = 4$ bits, and a large number of sampled values are simulated (10^5 samples). Fig. 3.1a shows that, as expected, the histogram is not uniformly distributed in all areas for s^n because $P(s^n = 0) = 0.6$ and $P(s^n = 1) = 0.4$. It can, however, be observed that a uniform histogram is obtained for x^n, and we have $P(x = 0) \approx P(x = 1) \approx 0.5$ for CBDS; see Fig. 3.1b. This first evaluation of the uniformity in CBDS is systematically verified using numerous tests presented in the next section.

3.4   TestU01 Statistical Test Results

The proposed CBDS has been evaluated using TestU01 for its statistical randomness. Table 3.1 lists seven batteries of tests in the TestU01 package. The "Standard" parameter in this table refers to the built-in parameters of the battery. The TestU01 suite implements 518 tests and reports p-values. If a p-value is within $[0.001, 0.999]$, the associated test is a success. A p-value lying outside this boundary means that its test has failed. The intention of CBDS is to enhance the effects of chaotic and random behaviors and to improve the statistical properties relative to s^n. This CBDS may utilize any reasonable random sequence as s^n. For demonstration purposes, ISAAC (indirection, shift, accumulate, add, and count) are adopted here. The x^n with this input can pass all of the performed tests in Table 3.1.

Table 3.1 TestU01 statistical test. (©Chinese Physical Society and IOP Publishing Ltd 2015, Reproduced with permission from [2])

Battery	Parameters	Number of statistics	ISAAC	CBDS
Rabbit	32×10^9 bits	40	0	0
Alphabit	32×10^9 bits	17	0	0
Pseudo DieHARD	Standard	126	0	0
FIPS_140_2	Standard	16	0	0
Small crush	Standard	15	0	0
Crush	Standard	144	0	0
Big crush	Standard	160	0	0
Numbers of failure		518	0	0

Fig. 3.2 The ring oscillator circuit schematic (with an odd number of NOT gates)

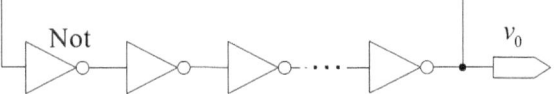

3.5   FPGA-Based Realization of CBDS

The design principle of CBDS is described and verified in this section.

The CBDS construction starts with the noise source design, which has been developed using ring oscillators. A ring oscillator is a device composed of an odd number of NOT gates as shown in Fig. 3.2, whose output oscillates between two voltage levels, representing true and false. As depicted, the output of the last inverter is fed back into the first one. Because a single inverter computes the logical NOT of its input, it can be deduced that the last output of a chain of an odd number of inverters is the logical NOT of the first input. This final output is asserted a finite amount of time after the first input is asserted, and thus the feedback of this last output to the first input causes oscillation.

A multiple ring-based design was developed with several ring oscillators with different ring lengths. Here we used three rings, each with long length gates. The three rings (as in Fig. 3.2) are XORed together to generate the output signal, as described in Fig. 3.3, where ring3:r31, ring3:r32, and ring3:r33 are our three long ring oscillators. The ring lengths for several ring oscillators were chosen in the range of relatively prime [5].

The ring oscillator output once XORed is sampled using the D flip-flop as shown in Fig. 3.3. The 'trigger' input has to be carried out from lower or similar frequencies of the 'clk'. This can cause more metastability effects in the D flip-flop, which reduces the mutual coupling effects. The output has to be subjected to postprocessing later, that is, to chaotic iterations.

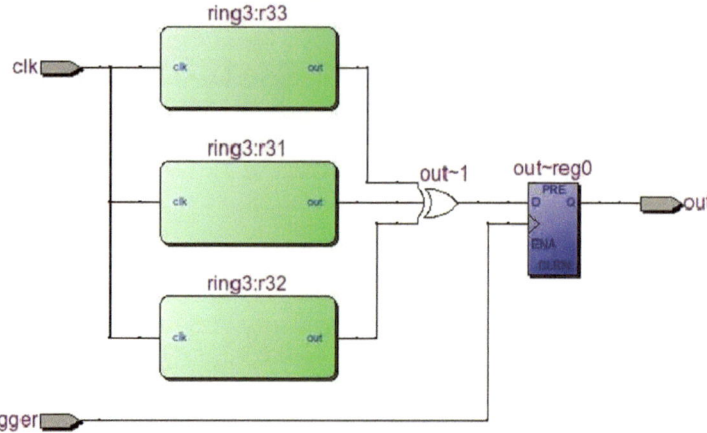

Fig. 3.3 A multiple ring-based design in FPGA. (©Chinese Physical Society and IOP Publishing Ltd 2015, Reproduced with permission from [2])

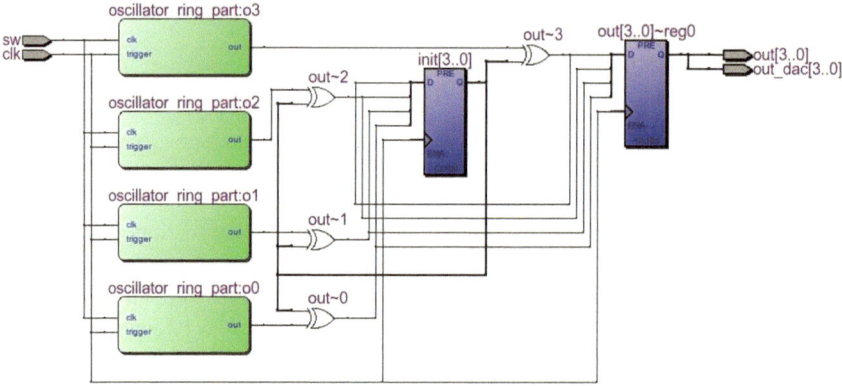

Fig. 3.4 The iterative equation design in FPGA. (©Chinese Physical Society and IOP Publishing Ltd 2015, Reproduced with permission from [2])

Each multiple ring oscillator outputs one component of state p^n (a binary digit). For our experiments, we have decided that p^n is constituted by four binary digits, which means that $N = 4$. The iterative equation is $x^n = x^{n-1} \oplus p^n$ as shown in Fig. 3.4. Two D flip-flops are used to store state information: init[3...0] is for x^0 whereas out[3...0] reg0 stores x^{n-1}.

The experimental observations of CBDS are finally shown in Fig. 3.5. Then a 10^8 bits long CBDS sequence is generated, and we have verified that it can pass ë the NIST statistical test suite.

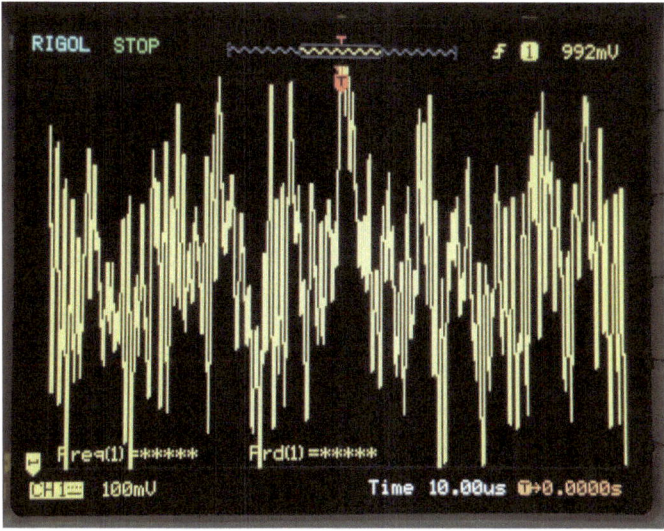

Fig. 3.5 The experimental observations of CBDS. (©Chinese Physical Society and IOP Publishing Ltd 2015, Reproduced with permission from [2])

References

1. Q. Wang, S. Yu, C. Guyeux, J. Bahi, X. Fang, Theoretical design and circuit implementation of integer domain chaotic systems. Int. J. Bifurc. Chaos **24**(10), 1450128-1 (2014). p. Art. no. 1450128
2. Q. Wang, S. Yu, C. Guyeux, J. Bahi, X. Fang, Study on a new chaotic bitwise dynamical system and its FPGA implementation. Chin. Phys. B **24**(6) (2015). art. no. 60503
3. Q. Wang, S. Yu, C. Li, J. Lü, X. Fang, C. Guyeux, J. Bahi, Theoretical design and FPGA-based implementation of higher-dimensional digital. IEEE Trans. Circuits Syst. I **63**(3), 401–412 (2016)
4. R. Couturier, C. Guyeux, *Designing Scientific Applications on GPUs* (CRC Press, Boca Raton, 2013)
5. B. Sunar, W. Martin, D. Stinson, A provably secure true random number generator with built-in tolerance to active attacks. IEEE Trans. Comput. **56**(1), 109–119 (2007)

Chapter 4
One-Dimensional Digital Chaotic Systems (ODDCS)

4.1 The Structure of One-Dimensional Digital Chaotic Systems

In this section, the structure of one-dimensional digital chaotic systems (ODDCS) in digital devices with finite precision is summarized, and the general framework of composing ODDCS is established.

4.1.1 The Conventional Iterative Update Mechanism

In traditional real domain chaotic systems (RDCS) studies, the general form of the iterative equation is:

$$x^n = f(x^{n-1}), \tag{4.1}$$

1. On the domain of infinite precision, the above Eq. (4.1) satisfies some definition of a chaotic system. Set the general binary form of the iteration value x^{n-1} and x^n as

$$\begin{cases} x^{n-1} = \ldots x_{P-1}^{n-1} x_{P-2}^{n-1} \ldots x_0^{n-1} . x_{-1}^{n-1} x_{-2}^{n-1} \ldots x_{-Q}^{n-1} \ldots, \\ x^n = \ldots x_{P-1}^n x_{P-2}^n \ldots x_0^n . x_{-1}^n x_{-2}^n \ldots x_{-Q}^n \ldots, \end{cases} \tag{4.2}$$

The general binary form of the iterative equation is represented as

$$\ldots x_{P-1}^n x_{P-2}^n \ldots x_0^n . x_{-1}^n \ldots x_{-Q}^n \ldots = f(\ldots x_{P-1}^{n-1} x_{P-2}^{n-1} \ldots x_0^{n-1} . x_{-1}^{n-1} \ldots x_{-Q}^{n-1} \ldots). \tag{4.3}$$

The above formula is further expressed as

$$\ldots x_{P-1}^n x_{P-2}^n \ldots x_0^n . x_{-1}^n x_{-2}^n \ldots x_{-Q}^n \ldots$$
$$= \ldots f(\cdot)_{P-1} f(\cdot)_{P-2} \ldots f(\cdot)_0 . f(\cdot)_{-1} \ldots f(\cdot)_{-Q} \ldots,$$

© The Author(s), under exclusive licence to Springer International Publishing AG, part of Springer Nature 2018
Q. Wang et al., *Design of Digital Chaotic Systems Updated by Random Iterations*, SpringerBriefs in Nonlinear Circuits, https://doi.org/10.1007/978-3-319-73549-8_4

where x_i^n ($i = \ldots, -2, -1, 0, 1, 2, \ldots$) denotes the ith bit of

$$\ldots x_{P-1}^n x_{P-2}^n \ldots x_0^n . x_{-1}^n x_{-2}^n \ldots x_{-Q}^n \ldots ,$$

$$f(\cdot) = f(\ldots x_{P-1}^{n-1} x_{P-2}^{n-1} \ldots x_0^{n-1} . x_{-1}^{n-1} x_{-2}^{n-1} \ldots x_{-Q}^{n-1} \ldots)$$

and $f(\cdot)_i$ denotes the ith bit of the result from the iterative equation $f(\cdot)$.

2. On the domain of N bits finite precision, the iterative equation outputs the iterative value as a periodic sequence, which can only be called "digital chaos." Set the general binary form of the iteration value x^{n-1} and x^n as

$$\begin{cases} x^{n-1} = x_{P-1}^{n-1} x_{P-2}^{n-1} \ldots x_0^{n-1} . x_{-1}^{n-1} x_{-2}^{n-1} \ldots x_{-Q}^{n-1} \\ x^n = x_{P-1}^n x_{P-2}^n \ldots x_0^n . x_{-1}^n x_{-2}^n \ldots x_{-Q}^n . \end{cases} \tag{4.4}$$

The general binary form of the iterative equation is represented as

$$x_{P-1}^n x_{P-2}^n \ldots x_0^n . x_{-1}^n x_{-2}^n \ldots x_{-Q}^n = f(x_{P-1}^{n-1} x_{P-2}^{n-1} \ldots x_0^{n-1} . x_{-1}^{n-1} \ldots x_{-Q}^{n-1}), \tag{4.5}$$

where $x_{P-1}^n x_{P-2}^n \ldots x_0^n$ and $x_{-1}^n x_{-2}^n \ldots x_{-Q}^n$ are, respectively, the binary representation of the integer part and the fractional part with $P + Q = N$. The above formula is further expressed as

$$(x_{P-1}^n x_{P-2}^n \ldots x_0^n . x_{-1}^n x_{-2}^n \ldots x_{-Q}^n)$$
$$= (f(\cdot)_{P-1} f(\cdot)_{P-2} \ldots f(\cdot)_0 . f(\cdot)_{-1} \ldots f(\cdot)_{-Q}),$$

where

$$f(\cdot)_i = f(x_{P-1}^{n-1} x_{P-2}^{n-1} \ldots x_0^{n-1} . x_{-1}^{n-1} x_{-2}^{n-1} \ldots x_{-Q}^{n-1})_i$$

denotes the ith bit of the result from the iterative equation $f(\cdot)$.

3. The main feature of the existing digital chaotic systems is that these systems are originally built on infinite domain, and then limited to finite domain. At each operation (iteration), all the bits in x^{n-1} will be updated by iterative equation f. Likewise, all the bits in x^n will be updated by iterative equation f at each operation (iteration). This is the main characteristic of the existing digital chaotic system and it is also called the iterative update mechanism.

4.1.2   The Iterative Update Mechanism Controlled by Random Sequences

Nowadays, the iterative update mechanism controlled by random sequences is used to build the chaotic systems. The main feature is that these systems are originally built on finite domain.

1. On a one-dimensional integer domain, let the iteration value $x_{N-1}^n x_{N-2}^n \ldots x_0^n$ and $x_{N-1}^{n-1} x_{N-2}^{n-1} \ldots x_0^{n-1}$ be binary integers in N-bit fixed precision, and the iterative equation be:

$$x_n = f(x_{n-1}) \rightarrow x_{N-1}^n x_{N-2}^n \ldots x_0^n = f(x_{N-1}^{n-1} x_{N-2}^{n-1} \ldots x_0^{n-1}). \qquad (4.6)$$

Then, the iterative equation with the conventional iterative update mechanism is

$$(x_{N-1}^n x_{N-2}^n \ldots x_0^n) = (f(\cdot)_{N-1} f(\cdot)_{N-2} \ldots f(\cdot)_0),$$

where $x_i^n (i = 0, 1, 2, \ldots, N-1)$ denotes the ith bit of $x_{N-1}^n x_{N-2}^n \ldots x_0^n$ and $f(\cdot)_i = f(x_{N-1}^{n-1} x_{N-2}^{n-1} \ldots x_0^{n-1})_i$ denote the ith bit of the result from the iterative equation $f(\cdot) = f(x_{N-1}^{n-1} x_{N-2}^{n-1} \ldots x_0^{n-1})$.

The iterative update mechanism controlled by random sequences is introduced; that is, through the random control sequence, some bits are randomly updated by f when the remaining bits keep the original value in each updating iteration. This is also referred to as a chaos generation strategy controlled by random sequences. Let the general expression of a one-sided infinite sequence be

$$s = s^1 s^2 \ldots s^n \ldots,$$

where s^1 is the first element of a random sequence, s^2 is the second one,..., s^n is the nth element of a random sequence, and so on.

Similarly, $s^n (n = 1, 2, 3, \ldots)$ is represented in binary form as

$$s^n = s_{N-1}^n s_{N-2}^n \ldots s_0^n,$$

where $s_j^n (j = 0, 1, 2, \ldots, N-1)$ denotes the jth bit of the random value. Note that the range of random numbers satisfies $s^n \in [0, 2^N - 1]$ due to the limited accuracy of N-bit representation.

Through the iterative update mechanism controlled by random sequences, the general form of the one-dimensional integer domain chaotic system is

$$x_j^n = \begin{cases} f(\cdot)_j & if \; s_j^n = 1 \\ x_j^{n-1} & if \; s_j^n = 0 \end{cases} (j = N - 1, N - 2, \ldots, 0), \qquad (4.7)$$

which is also called a chaos generation strategy controlled by random sequences for a one-dimensional integer domain chaotic system.

The above chaos generation shows that, if $s_j^n = 1 (j = N - 1, N - 2, \ldots, 0)$, then $x_j^n = f(\cdot)_j$ is obtained, which indicates that the jth bit of x can be updated by such an iterative equation. If $s_j^n = 0 (j = N - 1, N - 2, \ldots, 0)$, $x_j^n = x_j^{n-1}$ is obtained, which indicates that the jth bit of x remains unchanged. This is the most essential characteristic for the chaos generation strategy controlled by random sequences.

2. On the one-dimensional digital domain, set the iteration value

$$x^n_{P-1}x^n_{P-2}\ldots x^n_0.x^n_{-1}x^n_{-2}\ldots x^n_{-Q},$$

where $x^n_{P-1}x^n_{P-2}\ldots x^n_0$ and $x^n_{-1}x^n_{-2}\ldots x^n_{-Q}$ are, respectively, the binary representation of the integral part and the fractional one with $P + Q = N$.
Then the iterative equation with the conventional iterative update mechanism is

$$(x^n_{P-1}x^n_{P-2}\ldots x^n_0.x^n_{-1}x^n_{-2}\ldots x^n_{-Q})$$
$$= (f(\cdot)_{P-1}f(\cdot)_{P-2}\ldots f(\cdot)_0.f(\cdot)_{-1}f(\cdot)_{-2}\ldots f(\cdot)_{-Q}),$$

where $f(\cdot)_i = f(x^{n-1}_{P-1}x^{n-1}_{P-2}\ldots x^{n-1}_0.x^{n-1}_{-1}x^{n-1}_{-2}\ldots x^{n-1}_{-Q})_i$ denotes the ith bit of the result from the iterative equation $f(\cdot) = f(x^{n-1}_{P-1}x^{n-1}_{P-2}\ldots x^{n-1}_0.x^{n-1}_{-1}x^{n-1}_{-2}\ldots x^{n-1}_{-Q})$.
Let the general expression of a one-sided infinite sequence be

$$s = s^1 s^2 \ldots s^n \ldots,$$

where s^1 is the first element of the random sequence, s^2 is its second element, s^n is the nth element of the random sequence, and so on.
Similarly, $s^n (n = 1, 2, 3, \ldots)$ is represented in binary form as

$$s^n = s^n_{P-1}s^n_{P-2}\ldots s^n_0.s^n_{-1}s^n_{-2}\ldots s^n_{-Q},$$

where $s^n_j (j = P - 1, P - 2, \ldots, 0, -1, -2, \ldots, -Q)$ is the jth bit of the random value. Note that the range of random numbers satisfies $s^n \in [0, 2^P - 2^{-Q}]$ due to the limited accuracy of N-bit representation.
Through the iterative update mechanism controlled by random sequences, the general form of the one-dimensional integer domain chaotic system is

$$x^n_j = \begin{cases} f(\cdot)_j & if\ s^n_j = 1 \\ x^{n-1}_j & if\ s^n_j = 0 \end{cases} (j = P - 1, P - 2, \ldots, 0, -1, -2, \ldots, -Q),$$

(4.8)

which is also called the chaos generation strategy controlled by random sequences for a one-dimensional digital chaotic system.

4.2   The Connection Between a Chaotic System and Its Strongly Connected Network

In this section, we prove that ODDCS satisfies Devaney's definition of chaos if and only if the map of ODDCS is a strongly connected network [1]. IDCS and CBDS are two special cases in ODDCS.

4.2.1   Transitive Property of ODDCS

Proposition 4.1 *Function G_F is topological transitive in the metric space $(\mathcal{E}, d)$ if and only if its state network is strongly connected.*

Proof • First of all, the fact that a strongly connected network leads to topological transitivity for function G_f is proven.

As recalled previously, the so-called topological transitivity specifically refers to: for any nonempty open sets U_A and U_B in $(\mathcal{E}, d)$, there is always $n_0 > 0$ that makes $G_f^{n_0}(U_A) \cap U_B \neq \varnothing$.

Consider now two nonempty open sets U_A and U_B, and $(s_A, x_A) \in U_A$, $(s_B, x_B) \in U_B$. U_A and U_B are open, and we take a place in a metric space, thus there exist real numbers $r_A > 0$ and $r_B > 0$ such that the open ball $\mathcal{B}_A$ with center (s_A, x_A) and radius r_A is inside U_A (resp., the open ball $\mathcal{B}_b$ with center (s_B, x_B) and radius r_B is inside U_B). Without loss of generality, we can suppose that $r_A < 1$. We introduce the notations:

$$(s_A, x_A) = ((s_A^1 s_A^2 \ldots s_A^{n_0} \ldots s_A^n \ldots), x_A) \in U_A \subseteq \mathcal{E}$$

and

$$(s_B, x_B) = ((s_B^1 s_B^2 \ldots s_B^n \ldots), x_B) \in U_B \subseteq \mathcal{E}.$$

Let $(\tilde{s}, \tilde{x}) \in U_A$. If $\tilde{x} \neq x_A$, then $d_x(\tilde{x}, x_A) \geqslant 1$, and thus $d((s_A, x_A), (\tilde{s}, \tilde{x})) > 1$. Consequently, if $(\tilde{s}, \tilde{x}) \in \mathcal{B}_A$, then $d((s_A, x_A), (\tilde{s}, \tilde{x})) < r_A < 1$, and thus $\tilde{x} = x_A$. If we demand that the k_0 first elements of $\tilde{s}$ are equal to those from s_A, for the given r_A, an integer k_0 (i.e., a sequence $\tilde{s}$) can always be found to achieve $d_s(s_A, \tilde{s}) < r_A$. Suppose that, after the k_0th iteration, we have:

$$G_f^{k_0}(\tilde{s}, \tilde{x}) = (s_C, x_C).$$

Inasmuch as G_f is strongly connected, there is at least one path from the state x_C to the state x_B after other i_0 iterations. Because

$$G_f^{i_0}(s_C, x_C) = (s_B, x_B)$$

holds, $n_0 = k_0 + i_0$ is found,
therefore the point $(\tilde{s}, \tilde{x}) \in U_A$ satisfies

$$G_f^{n_0}(\tilde{s}, \tilde{x}) = (s_B, x_B) \in G_f^{n_0}(U_A) \cap U_B$$

with $n_0 = k_0 + i_0$, making the claim

$$G_f^{n_0}(U_A) \cap U_B \neq \varnothing$$

true.

- Then we only need to prove that, if the state network is not strongly connected, then G_f is not transitive.

  Assuming there is no path from the vertex x_A to the vertex x_B, then the state network for function G_f is not a strongly connected graph, and for all points $(\tilde{s}, \tilde{x})$ that belong to U_A, the x component of the points must be $\tilde{x} = x_A$ to satisfy $d((s_A, x_A), (\tilde{s}, \tilde{x})) < r_A. \forall k, (G_f^k(\tilde{s}, x_A))_x \neq x_B$, because there is no path from the vertex x_A to the vertex x_B. That means $\forall k, (G_f^k(\tilde{s}, x_A))_x \notin U_B$ then $\forall k, G_f^k(U_A) \cap U_B = \varnothing$, therefore G_f is not transitive.

4.2.2 Dense Periodic Points of ODDCS

Proposition 4.2 *If the state network is strongly connected, then its periodic points are dense.*

Proof We want to show that, for any given $\varepsilon > 0$, a periodic point $(\tilde{s}, \tilde{x}) \in \mathscr{E}$ can always be found within the range ε of any point $(\hat{s}, \hat{x}) \in \mathscr{E}$. Without loss of generality, we assume that the given $\varepsilon < 1$ and that the general form of $(\hat{s}, \hat{x})$ is

$$(\hat{s}, \hat{x}) = ((\hat{s}^1 \hat{s}^2 \dots \hat{s}^{k_0} \dots \hat{s}^n \dots), \hat{x}) \in \mathscr{E}$$

If $\tilde{x} \neq \hat{x}$, then $d_x(\tilde{x}, \hat{x}) \geqslant 1$, and therefore $d((\hat{s}, \hat{x}), (\tilde{s}, \tilde{x})) > 1$. Thus $\tilde{x} = \hat{x}$. If the k_0 first elements of $\hat{s}$ and $\tilde{s}$ are the same, then $d_s(\hat{s}, \tilde{s}) < \varepsilon$.

If after the k_0th iteration, we have

$$(G_f^{k_0}(\tilde{s}, \tilde{x}))_x = x';$$

then, because G_f is strongly connected, there is at least one path from the state x' to the state $\hat{x}$ after other i_0 iterations. Because

$$G_f^{i_0}(s', x') = (\tilde{s}, \tilde{x})$$

holds, $n_0 = k_0 + i_0$ is found, and after another i_0 iterations, the path for s is $\tilde{s}^{k_0+1} \dots \tilde{s}^{k_0+i_0}$, and the following condition is met

$$\tilde{x} = \hat{x} = (G_f^{k_0+i_0}(\tilde{s}, \tilde{x}))_x.$$

Then, within the range ε of the point $(\hat{s}, \hat{x})$, one can find the periodic point:

$$(\tilde{s}, \tilde{x}) = ((\hat{s}^1 \hat{s}^2 \dots \hat{s}^{k_0} \tilde{s}^{k_0+1} \dots \tilde{s}^{k_0+i_0} \dots \hat{s}^1 \hat{s}^2 \dots \hat{s}^{k_0} \tilde{s}^{k_0+1} \dots \tilde{s}^{k_0+i_0} \dots), \hat{x})$$

making

$$d((\hat{s}, \hat{x}), (\tilde{s}, \tilde{x})) = d_s(\hat{s}, \tilde{s}) + d_x(\hat{x}, \tilde{x}) = d_s(\hat{s}, \tilde{s}) < \varepsilon$$

true.

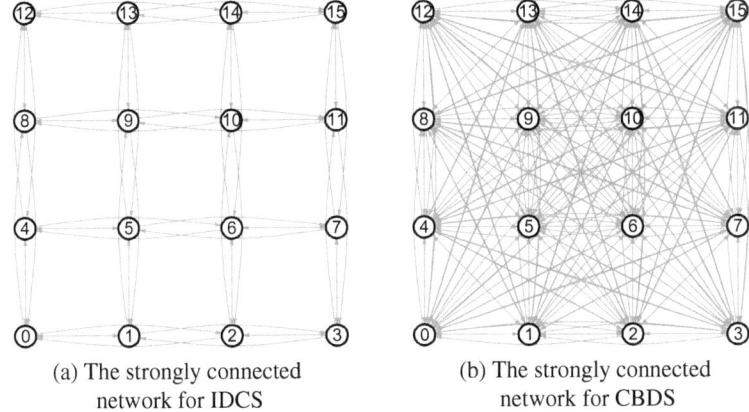

(a) The strongly connected network for IDCS

(b) The strongly connected network for CBDS

Fig. 4.1 The strongly connected network

In summary, the periodic points of G_f are dense in $\mathscr{E}$.

4.2.3 Chaotic System and Its Strongly Connected Network

Because IDCS and CBDS process data on the integer domain, they are ODDCS with $N = P$ ($Q = 0$). Considering each state of the system as a vertex, and the mapping relation as a directed edge, the state network of the chaotic system can be built up. As shown in Fig. 4.1a and b, the state networks of IDCS and CBDS with $N = 4$ are strongly connected, therefore their chaotic behavior has been reconfirmed.

4.3 Lyapunov Exponents of a Class of ODDCS

In this section, the Lyapunov exponents of ODDCS with $N = P$ ($Q = 0$) are estimated.

4.3.1 General Expression of Equivalent Decimal for G_F

Set the binary form of N-bit integers as $x = x_{N-1}x_{N-2}\ldots x_0$, where $x_i \in \{0, 1\}$ ($i = N - 1, N - 2, \ldots, 0$).

The general form of the corresponding decimal integer is obtained as

$$X = \sum_{k=0}^{N-1}(x_k \cdot 2^k). \tag{4.9}$$

The random number of the s sequence is expressed in the corresponding decimal fraction as

$$S = \sum_{k=1}^{+\infty}(s^k \cdot 2^{-Nk}), \tag{4.10}$$

where $2^{-Nk}(k = 1, 2, \ldots)$ denote weights.

According to Eqs. (4.9) and (4.10) added together, the general form of the corresponding decimal number is obtained as

$$y = \sum_{k=0}^{N-1}(x_k \cdot 2^k) + \sum_{k=1}^{+\infty}(s^k \cdot 2^{-Nk}).$$

According to the chaos generation strategy controlled by random sequences, the general form of a one-dimensional digital discrete-time iterative equation can be presented as

$$g(X) = (x \cdot \overline{s^1}) + (f(x) \cdot s^1). \tag{4.11}$$

Separately shifting one value in each one-sided infinite sequence $(s = s^1 s^2 \ldots s^n \ldots)$, the first value turns into s^2 and the corresponding weight is 2^{-N}. Thus the general form of the corresponding decimal fraction after shifting one value in every one-sided infinite sequence is

$$g(S) = 2^N \sum_{k=2}^{+\infty}(s^k \cdot 2^{-Nk}). \tag{4.12}$$

Adding Eqs. (4.11) and (4.12) together, one can obtain the general form of the corresponding decimal number,

$$g(y) = (x \cdot \overline{s^1}) + (f(x) \cdot s^1) + 2^N \sum_{k=2}^{+\infty}(s^k \cdot 2^{-Nk}),$$

after randomly updating multiple random bits and shifting one value in every one-sided infinite sequence.

4.3.2 *Mathematical Expression for* $\frac{\partial G(y)}{\partial y}$

In the interval $[\frac{n}{2^N}, \frac{n+1}{2^N})$ $(n \in [0, 2^{2N} - 1])$, the decimal integer part is not changed where $\Delta X = 0$. Furthermore, the first decimals of the m sequences are the same, therefore

$$\Delta y = \Delta X + \Delta S = \Delta S = \sum_{k=2}^{+\infty}(\Delta s^k \cdot 2^{-Nk}),$$

From the definition of the partial derivative, one has

$$
\begin{aligned}
\frac{\partial g(y)}{\partial y} &= \lim_{\Delta y \to 0} \frac{g(y+\Delta y)-g(y)}{\Delta y} \\
&= \lim_{\Delta S \to 0} \left(\frac{g(X)+2^N \sum_{k=2}^{+\infty}((s^k+\Delta s^k)\cdot 2^{-Nk})}{\sum_{k=2}^{+\infty}(\Delta s^k)\cdot 2^{-Nk}} - \frac{g(X)+2^N \sum_{k=2}^{+\infty}(s^k \cdot 2^{-Nk})}{\sum_{k=2}^{+\infty}(\Delta s^k)\cdot 2^{-Nk}} \right) \\
&= \lim_{\Delta S \to 0} \frac{2^N \sum_{k=2}^{+\infty}(\Delta s^k \cdot 2^{-Nk})}{\sum_{k=2}^{+\infty}(\Delta s^k \cdot 2^{-Nk})} \\
&= 2^N .
\end{aligned}
$$

4.3.3 Estimating the Lyapunov Exponents

The Lyapunov exponent of the specific ODDCS can be estimated as

$$
\begin{aligned}
\lambda(y) &= \lim_{n \to +\infty} \frac{1}{2n} \sum_{i=1}^{n} \ln |g'(y^{i-1})| \\
&= \lim_{n \to +\infty} \frac{1}{2n} \ln((2^N)^{2n}) \\
&= N \ln 2,
\end{aligned}
\tag{4.13}
$$

Some concrete examples are provided to illustrate the parameters in Eq. (4.13).

1. Set $N = 4$, the relationship between y and X is shown in Fig. 4.2a, thus any two points in the interval $[\frac{n}{4}, \frac{n+1}{4})(n \in \{0, 1, \dots, 2^4\})$ have the same integer part. Similarly, set $N = 10$; the relationship between y and X is shown in Fig. 4.2b, thus any two points in the interval $[\frac{n}{10}, \frac{n+1}{10})(n \in \{0, 1, \dots, 2^{10}\})$ have the same integer part. In the general case N, any two points in the interval $[\frac{n}{N}, \frac{n+1}{N})(n \in \{0, 1, \dots, 2^N\}, N = 2, 3, \dots)$ have the same integer part.
2. According to Eq. (4.13), the corresponding waveform for $N = 4$ is shown in Fig. 4.3a. Obviously, except for the break point, the slope of the remaining part is 4, then $\frac{dg(y)}{dy} = 4$ is established.
3. Set $N = 10$; the corresponding waveform is shown in Fig. 4.3b. Obviously, except for the break point, the slope of the remaining part is 10; then $\frac{dg(y)}{dy} = 10$ is established.
4. In general, Lyapunov exponents for RDCS can be positive, negative, or zero. Positive Lyapunov exponents indicate divergence from initial conditions, especially the bounded system with positive Lyapunov exponents is chaotic. Negative values for the exponents indicate convergence to a fixed point, and zero indicates neither divergence nor convergence; that is, one is usually faced with periodic motion. According to Eq. (4.13), a dynamical system with $N > 1$, positive Lyapunov exponents $\lambda(y) > 0$, and global bounded movement $|x^n| \le 2^N - 1 (n = 0, 1, 2, \dots)$, displays chaotic behavior. For example, set $N = 4$ and Lyapunov exponents $\lambda(y) > 0$; then the simulated chaotic waveform is shown in Fig. 4.4a.

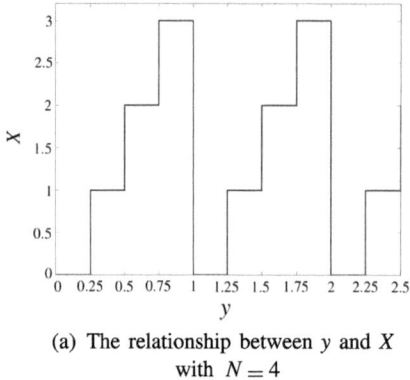

(a) The relationship between y and X with $N = 4$

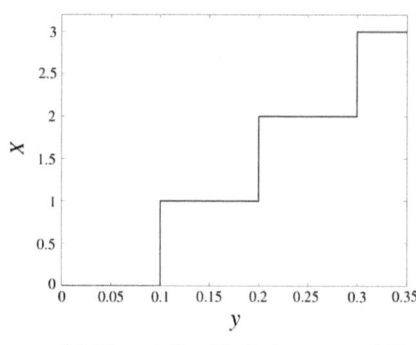

(b) The relationship between y and X with $N = 10$

Fig. 4.2   The relationship between y and X

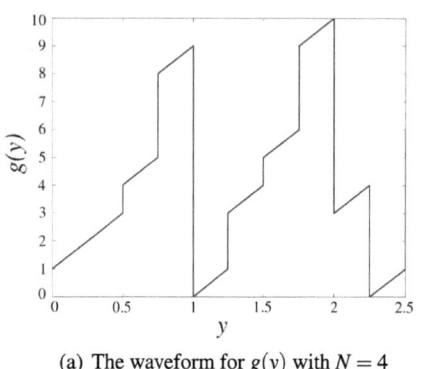

(a) The waveform for $g(y)$ with $N = 4$

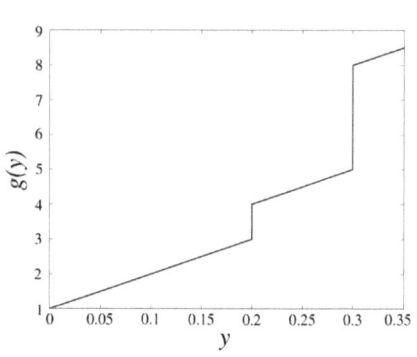

(b) The waveform for $g(y)$ with $N = 10$

Fig. 4.3   The waveform for $g(y)$

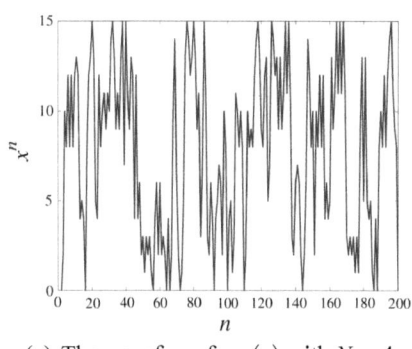

(a) The waveform for $g(y)$ with $N = 4$

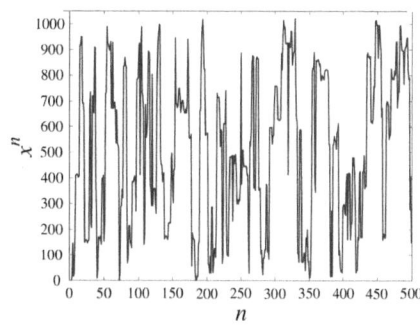

(b) The waveform for $g(y)$ with $N = 10$

Fig. 4.4   The waveform for $g(y)$

5. According to Eq. (4.13), the chaotic property is stronger as the Lyapunov exponent increases with the number N. Therefore, a larger N for an integer domain chaotic system is usually recommended in practical applications. The simulated chaotic waveform with $N = 10$ is shown in Fig. 4.4b for comparison.

Reference

1. J. Bahi, J.F. Couchot, C. Guyeux, A. Richard, On the link between strongly connected iteration graphs and chaotic boolean discrete-time dynamical systems, in *International Symposium on Fundamentals of Computation Theory* (Springer, Berlin, 2011), pp. 126–137

Chapter 5
Higher-Dimensional Digital Chaotic Systems (HDDCS)

5.1 Design of HDDCS

5.1.1 Higher-Dimensional Integer Domain Chaotic Systems (HDDCS)

On a m-dimensional integer domain, set one of the iteration value

$$x^n_{k,N-1}x^n_{k,N-2}\cdots x^n_{k,0}.$$

Then, the iterative equation with the conventional iterative update mechanism is

$$\begin{cases} x^n_{1,N-1}x^n_{1,N-2}\cdots x^n_{1,0} = F_1(\cdot)_{N-1}\ldots F_1(\cdot)_0, \\ x^n_{2,N-1}x^n_{2,N-2}\cdots x^n_{2,0} = F_2(\cdot)_{N-1}\ldots F_2(\cdot)_0, \\ \quad\vdots \\ x^n_{m,N-1}x^n_{m,N-2}\cdots x^n_{m,0} = F_m(\cdot)_{N-1}\ldots F_m(\cdot)_0, \end{cases} \tag{5.1}$$

where

$$\begin{cases} F_1(\cdot)_i = F_1(x^{n-1}_1, x^{n-1}_2, \ldots, x^{n-1}_m)_i, \\ F_2(\cdot)_i = F_2(x^{n-1}_1, x^{n-1}_2, \ldots, x^{n-1}_m)_i, \\ \quad\vdots \\ F_m(\cdot)_i = F_m(x^{n-1}_1, x^{n-1}_2, \ldots, x^{n-1}_m)_i. \end{cases}$$

denotes the ith component of the iterative function, where $i = N-1, N-2, \ldots, 0$.

Parts of this chapter were reproduced with permission from [5] ©IEEE 2016, [6] ©World Scientific Publishing Co Pte Ltd 2014, and [7] ©Chinese Physical Society and IOP Publishing Ltd 2015.

Q. Wang et al., *Design of Digital Chaotic Systems Updated by Random Iterations*, SpringerBriefs in Nonlinear Circuits, https://doi.org/10.1007/978-3-319-73549-8_5

Let the general expression of m one-sided infinite random sequences be

$$
\begin{cases}
s = s^1 s^2 \dots s^n \dots, \\
u = u^1 u^2 \dots u^n \dots, \\
\quad \vdots \\
v = v^1 v^2 \dots v^n \dots.
\end{cases}
\tag{5.2}
$$

Similarly, each random number of these m sequences is represented in binary form as

$$
\begin{cases}
s^n = s^n_{N-1} s^n_{N-2} \cdots s^n_0, \\
u^n = u^n_{N-1} u^n_{N-2} \cdots u^n_0, \\
\quad \vdots \\
v^n = v^n_{N-1} v^n_{N-2} \cdots v^n_0,
\end{cases}
$$

where $n \in \mathbb{Z}^+, \forall j \in \{N-1, N-2, \dots, 0\}$, s^n_j, u^n_j, and v^n_j are the jth bit of the binary form of s^n, u^n, and v^n, respectively. Note that the range of random numbers satisfies $s^n, u^n, \dots, v^n \in [0, 2^{N-1}]$ due to the limited accuracy of N-bit representation.

Through the iterative update mechanism controlled by random sequences, the general form of the m-dimensional integer domain chaotic system is

$$
\begin{cases}
x^n_{1,N-1} x^n_{1,N-2} \cdots x^n_{1,0} = F_1(\cdot)_{s^n_{N-1}} \cdots F_1(\cdot)_{s^n_0}, \\
x^n_{2,N-1} x^n_{2,N-2} \cdots x^n_{2,0} = F_2(\cdot)_{u^n_{P-1}} \cdots F_2(\cdot)_{u^n_0}, \\
\quad \vdots \\
x^n_{m,N-1} x^n_{m,N-2} \cdots x^n_{m,0} = F_m(\cdot)_{v^n_{P-1}} \cdots F_m(\cdot)_{v^n_0},
\end{cases}
\tag{5.3}
$$

where

$$
\begin{cases}
F_1(\cdot)_i = F_1(x^{n-1}_1, x^{n-1}_2, \dots, x^{n-1}_m)_i \\
F_2(\cdot)_i = F_2(x^{n-1}_1, x^{n-1}_2, \dots, x^{n-1}_m)_i \\
\quad \vdots \\
F_m(\cdot)_i = F_m(x^{n-1}_1, x^{n-1}_2, \dots, x^{n-1}_m)_i
\end{cases}
$$

denotes the ith component of the iterative function, where $i = N-1, N-2, \dots, 0$.

In Eq. (5.3), we define the chaos generation strategy controlled by random sequences for HDDCS as

$$
\begin{cases}
x_{1,j}^n = F_1(\cdot)_{s_j^n} = \begin{cases} F_1(\cdot)_j & \text{if } s_j^n = 1, \\ x_{1,j}^{n-1} & \text{if } s_j^n = 0, \end{cases} \\[2ex]
x_{2,j}^n = F_2(\cdot)_{u_j^n} = \begin{cases} F_2(\cdot)_j & \text{if } u_j^n = 1, \\ x_{2,j}^{n-1} & \text{if } u_j^n = 0, \end{cases} \\[2ex]
\qquad\vdots \\[1ex]
x_{m,j}^n = F_m(\cdot)_{v_j^n} = \begin{cases} F_m(\cdot)_j & \text{if } v_j^n = 1, \\ x_{m,j}^{n-1} & \text{if } v_j^n = 0, \end{cases}
\end{cases}
\tag{5.4}
$$

where $j = N - 1, N - 2, \ldots, 0$.

5.1.2 Description of HDDCS

In HDDCS, the range of the input data is extended to the digital domain of the finite integer part and the finite fractional part. The general form of the iterative equation for an m-dimensional digital system with N-bit fixed precision is

$$
\begin{cases}
x_1^n = F_1(x_1^{n-1}, x_2^{n-1}, \ldots, x_m^{n-1}), \\
x_2^n = F_2(x_1^{n-1}, x_2^{n-1}, \ldots, x_m^{n-1}), \\
\quad\vdots \\
x_m^n = F_m(x_1^{n-1}, x_2^{n-1}, \ldots, x_m^{n-1}),
\end{cases}
$$

where $F_1, F_2, \cdots, F_m$ are iterative functions, $x_1^n, x_2^n, \cdots, x_m^n$ and $x_1^{n-1}, x_2^{n-1}, \cdots, x_m^{n-1}$ can be represented as the binary form:

$$
\begin{cases}
x_1^n = x_{1,P-1}^n x_{1,P-2}^n \cdots x_{1,0}^n . x_{1,-1}^n x_{1,-2}^n \cdots x_{1,-Q}^n, \\
x_1^{n-1} = x_{1,P-1}^{n-1} x_{1,P-2}^{n-1} \cdots x_{1,0}^{n-1} . x_{1,-1}^{n-1} x_{1,-2}^{n-1} \cdots x_{1,-Q}^{n-1}, \\
x_2^n = x_{2,P-1}^n x_{2,P-2}^n \cdots x_{2,0}^n . x_{2,-1}^n x_{2,-2}^n \cdots x_{2,-Q}^n, \\
x_2^{n-1} = x_{2,P-1}^{n-1} x_{2,P-2}^{n-1} \cdots x_{2,0}^{n-1} . x_{2,-1}^{n-1} x_{2,-2}^{n-1} \cdots x_{2,-Q}^{n-1}, \\
\quad\vdots \\
x_m^n = x_{m,P-1}^n x_{m,P-2}^n \cdots x_{m,0}^n . x_{m,-1}^n x_{m,-2}^n \cdots x_{m,-Q}^n, \\
x_m^{n-1} = x_{m,P-1}^{n-1} x_{m,P-2}^{n-1} \cdots x_{m,0}^{n-1} . x_{m,-1}^{n-1} x_{m,-2}^{n-1} \cdots x_{m,-Q}^{n-1},
\end{cases}
\tag{5.5}
$$

and $P + Q = N$.

Set the general expression of m one-sided infinite random sequences as

$$
\begin{cases}
s = s^1 s^2 \dots s^n \dots, \\
u = u^1 u^2 \dots u^n \dots, \\
\quad \vdots \\
v = v^1 v^2 \dots v^n \dots.
\end{cases}
\tag{5.6}
$$

Likewise, each random number of these m sequences is expressed in binary form as

$$
\begin{cases}
s^n = s^n_{P-1} s^n_{P-2} \dots s^n_0 . s^n_{-1} s^n_{-2} \dots s^n_{-Q}, \\
u^n = u^n_{P-1} u^n_{P-2} \dots u^n_0 . u^n_{-1} u^n_{-2} \dots u^n_{-Q}, \\
\quad \vdots \\
v^n = v^n_{P-1} v^n_{P-2} \dots v^n_0 . v^n_{-1} v^n_{-2} \dots v^n_{-Q},
\end{cases}
$$

where $n \in \mathbb{Z}^+, \forall \, j \in \{P-1, P-2, \dots, 0, -1, -2, \dots, -Q\}$, s^n_j, u^n_j, and v^n_j are the jth bits of the binary form of s^n, u^n, and v^n, respectively. Note that the range of random numbers satisfies $s^n, u^n, \dots, v^n \in [0, 2^P - 2^{-Q}]$, again due to the limited accuracy of N-bit representation.

Through the iterative update mechanism controlled by random sequences, the general form of an m-dimensional digital chaotic system (m-DDCS) is

$$
\begin{cases}
x^n_{1,P-1} x^n_{1,P-2} \dots x^n_{1,0} . x^n_{1,-1} x^n_{1,-2} \dots x^n_{1,-Q} \\
\quad = F_1(\cdot)_{s^n_{P-1}} \dots F_1(\cdot)_{s^n_0} . F_1(\cdot)_{s^n_{-1}} F_1(\cdot)_{s^n_{-2}} \dots F_1(\cdot)_{s^n_{-Q}}, \\
x^n_{2,P-1} x^n_{2,P-2} \dots x^n_{2,0} . x^n_{2,-1} x^n_{2,-2} \dots x^n_{2,-Q} \\
\quad = F_2(\cdot)_{u^n_{P-1}} \dots F_2(\cdot)_{u^n_0} . F_2(\cdot)_{u^n_{-1}} F_2(\cdot)_{u^n_{-2}} \dots F_2(\cdot)_{u^n_{-Q}}, \\
\quad \vdots \\
x^n_{m,P-1} x^n_{m,P-2} \dots x^n_{m,0} . x^n_{m,-1} x^n_{m,-2} \dots x^n_{m,-Q} \\
\quad = F_m(\cdot)_{v^n_{P-1}} \dots F_m(\cdot)_{v^n_0} . F_m(\cdot)_{v^n_{-1}} F_m(\cdot)_{v^n_{-2}} \dots F_m(\cdot)_{v^n_{-Q}},
\end{cases}
\tag{5.7}
$$

where

$$
\begin{cases}
F_1(\cdot)_i = F_1(x^{n-1}_1, x^{n-1}_2, \dots, x^{n-1}_m)_i \\
F_2(\cdot)_i = F_2(x^{n-1}_1, x^{n-1}_2, \dots, x^{n-1}_m)_i \\
\quad \vdots \\
F_m(\cdot)_i = F_m(x^{n-1}_1, x^{n-1}_2, \dots, x^{n-1}_m)_i
\end{cases}
$$

stands for the ith component of the iterative function, where $i = P-1, P-2, \dots, 0, -1, -2, \dots, -Q$.

In Eq. (5.7), we define the chaos generation strategy controlled by random sequences for HDDCS as

$$
\begin{cases}
x_{1,j}^n = F_1(\cdot)_{s_j^n} = \begin{cases} F_1(\cdot)_j & \text{if } s_j^n = 1, \\ x_{1,j}^{n-1} & \text{if } s_j^n = 0, \end{cases} \\[2mm]
x_{2,j}^n = F_2(\cdot)_{u_j^n} = \begin{cases} F_2(\cdot)_j & \text{if } u_j^n = 1, \\ x_{2,j}^{n-1} & \text{if } u_j^n = 0, \end{cases} \\[2mm]
\quad\vdots \\[2mm]
x_{m,j}^n = F_m(\cdot)_{v_j^n} = \begin{cases} F_m(\cdot)_j & \text{if } v_j^n = 1, \\ x_{m,j}^{n-1} & \text{if } v_j^n = 0, \end{cases}
\end{cases}
\tag{5.8}
$$

where $j = P - 1, P - 2, \ldots, 0, -1, -2, \ldots, -Q$.

Then, Eq. (5.7) is further expressed as the general form:

$$
\begin{cases}
x_1^n = (x_1^{n-1} \cdot \overline{s^n}) + (F_1(\cdot) \cdot s^n), \\
x_2^n = (x_2^{n-1} \cdot \overline{u^n}) + (F_2(\cdot) \cdot u^n), \\
\quad\vdots \\
x_m^n = (x_m^{n-1} \cdot \overline{v^n}) + (F_m(\cdot) \cdot v^n),
\end{cases}
\tag{5.9}
$$

where the operators ".", "$\overline{(\cdot)}$", and "+" denote bitwise AND, bitwise NOT (negation), and bitwise OR, respectively.

Let us define $\mathscr{E}$ as the set of points E of the form $((s, u, \ldots, v), (x_1, x_2, \ldots, x_m))$, where $s, u, \ldots, v$ are m independent random sequences and $x_1, x_2, \ldots, x_m$ are N-bit real numbers. Consider a metric space $(\mathscr{E}, d)$ and a continuous function $G_F : \mathscr{E} \to \mathscr{E}$ as

$$
\begin{aligned}
G_F(E) &= G_F((s, u, \ldots, v), (x_1, x_2, \ldots, x_m)) \\
&= ((\sigma(s), \sigma(u), \ldots, \sigma(v)), (H_{F_1}(i(s), (x_1, x_2, \ldots, \\
&\quad x_m)), H_{F_2}(i(u), (x_1, x_2, \ldots, x_m)), \ldots, H_{F_m}(i(v), \\
&\quad (x_1, x_2, \ldots, x_m))),
\end{aligned}
\tag{5.10}
$$

where $\sigma(w)$ ($w \in \{s, u, \cdots, v\}$) shifts one integer in the one-sided infinite sequence $w = (w^1 w^2 \ldots w^n \ldots)$ to the left, and

$$
\begin{cases}
H_{F_1}(i(s), (x_1, x_2, \ldots, x_m)) = ((x_1 \cdot \overline{i(s)}) + (F_1(\cdot) \cdot i(s))), \\
H_{F_2}(i(u), (x_1, x_2, \ldots, x_m)) = ((x_2 \cdot \overline{i(u)}) + (F_2(\cdot) \cdot i(u))), \\
\quad\vdots \\
H_{F_m}(i(v), (x_1, x_2, \ldots, x_m)) = ((x_m \cdot \overline{i(v)}) + (F_m(\cdot) \cdot i(v))).
\end{cases}
\tag{5.11}
$$

As for function $\sigma(w)$, one has

$$\sigma^k(w) = w^{k+1}w^{k+2}\dots w^n,$$

where k is a positive integer, and

$$\sigma^k(w) \triangleq \underbrace{\sigma \circ \sigma \circ \dots \circ \sigma}_{k}(w).$$

Then, one can further get

$$i(w) = w^k,$$

where $w \in \{s, u, \cdots, v\}$, $k \in \mathbb{Z}^+$, and $i(w)$ is equal to the overflow from the left shifting of the sequence w. In other words, some bits are randomly updated by F_1 in each updating iteration, which are determined by $i(s)$. Similarly, $i(u)$ determines the specific bits that are updated by F_2, and $i(v)$ determines the specific bits that are updated by F_m. The relation among the m-dimensional digital domain system, m random sequences, and m-dimensional digital chaotic system is shown in Fig. 5.1. Note that true random sequences usually cannot be obtained via digital simulation [1, 2]. But the input m random sequences can be generated by a true random number generator (TRNG) based on digital circuit artifacts, such as metastable circuits [3], and digital oscillator rings [4].

Let $E^0 = ((s, u, \dots, v), (x_1^0, x_2^0, \dots, x_m^0)) \in \mathscr{E}$ be the initial condition,

$$E^k = ((\sigma^k(s), \sigma^k(u), \cdots, \sigma^k(v))), (x_1^k, x_2^k, \cdots, x_m^k)) \in \mathscr{E},$$

and

$$E^{k+1} = ((\sigma^{k+1}(s), \sigma^{k+1}(u), \cdots, \sigma^{k+1}(v)), (x_1^{k+1}, x_2^{k+1}, \cdots, x_m^{k+1})) \in \mathscr{E}$$

denote the kth and $(k+1)$th iteration, respectively.

Fig. 5.1 The flowchart of the proposed higher-dimensional digital chaotic system. (©IEEE 2016, Reproduced with permission from [5])

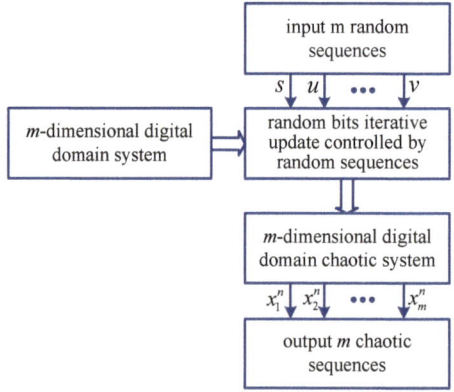

With the above notations, HDDCS can be defined as

$$E^{k+1} = G_F(E^k),$$

where $k = 0, 1, 2, 3, \ldots$.

Let

$$\begin{cases} \hat{s} = \hat{s}^1 \hat{s}^2 \ldots \hat{s}^n \ldots \\ \hat{u} = \hat{u}^1 \hat{u}^2 \ldots \hat{u}^n \ldots \\ \vdots \\ \hat{v} = \hat{v}^1 \hat{v}^2 \ldots \hat{v}^n \ldots \end{cases}$$

denote another one-sided infinite random sequence. Then, we introduce the distance d in the metric space $(\mathcal{E}, d)$ as

$$d(((s, u, \ldots, v), (x_1, x_2, \cdots, x_m)), ((\hat{s}, \hat{u}, \ldots, \hat{v}), (\hat{x}_1, \hat{x}_2, \ldots, \hat{x}_m)))$$
$$= d_s(s, \hat{s}) + d_u(u, \hat{u}) + \cdots + d_v(v, \hat{v}) + d_x((x_1, x_2, \cdots, x_m), (\hat{x}_1, \hat{x}_2, \cdots, \hat{x}_m)),$$

where

$$\begin{cases} d_s(s, \hat{s}) = \sum_{k=1}^{\infty} \frac{|s^k - \hat{s}^k|}{2^{Nk}}, \\ d_u(u, \hat{u}) = \sum_{k=1}^{\infty} \frac{|u^k - \hat{u}^k|}{2^{Nk}}, \\ \vdots \\ d_v(v, \hat{v}) = \sum_{k=1}^{\infty} \frac{|v^k - \hat{v}^k|}{2^{Nk}}, \end{cases}$$

$$d_x((x_1, x_2, \cdots, x_m), (\hat{x}_1, \hat{x}_2, \cdots, \hat{x}_m))$$
$$= \sqrt{(x_1 - \hat{x}_1)^2 + (x_2 - \hat{x}_2)^2 + \ldots + (x_m - \hat{x}_m)^2},$$

$x_1, \hat{x}_1, x_2, \hat{x}_2, \ldots, x_m, \hat{x}_m$ are binary forms of real numbers with N-bit finite precision, and $0 \leq d_x \leq \sqrt{m}(2^P - 2^{-Q}))$.

5.1.3 Comparative Study of RDCS, IDCS, CBDS, and HDDCS

The proposed HDDCS system is briefly compared with RDCS and IDCS in [6], CBDS in [7], and in Table 5.1. More detailed differences are summarized below.

- The adopted implementation environment of RDCS is an infinite precision. However, IDCS, CBDS, and HDDCS are implemented in finite precision; they are suitable for digital computers or other digital devices.

Table 5.1 Comparison of the digital chaotic systems, RDCS, IDCS, CBDS, and HDDCS. (©IEEE 2016, Reproduced with permission from [5])

Name	RDCS	IDCS	CBDS	HDDCS
Domain	Real	Integer		Digital
Precision	Infinite	Finite (N bits)		Finite ($N = P + Q$, P bits integer, Q bits fraction)
Dimension	m-D	1-D		m-D
External control inputs	Null	$s = s^1 s^2 \cdots s^n \cdots$ $s^n \in \{0, 1, 2, \cdots, N-1\}$	$s = s^1 s^2 \cdots s^n \cdots$ $s^n = (s^n_{N-1} s^n_{N-2} \cdots s^n_0)_2$ $s^n \in [0, 2^N - 1]$	$\left.\begin{array}{l} s = s^1 s^2 \cdots s^n \cdots \\ u = u^1 u^2 \cdots u^n \cdots \\ \vdots \\ v = v^1 v^2 \cdots v^n \cdots \end{array}\right\}$ $\left.\begin{array}{l} s^n = (s^n_{P-1} s^n_{P-2} \cdots s^n_0 \cdot s^2_{-1} s^n_{-2} \cdots s^n_{-Q})_2 \\ u^n = (u^n_{P-1} u^n_{P-2} \cdots u^n_0 \cdot u^2_{-1} u^n_{-2} \cdots u^n_{-Q})_2 \\ \vdots \\ v^n = (v^n_{P-1} v^n_{P-2} \cdots v^n_0 \cdot v^2_{-1} v^n_{-2} \cdots v^n_{-Q})_2 \end{array}\right.$ $s^n, u^n, \cdots, v^n \in [0, 2^P - 2^{-Q}]$

(continued)

Table 5.1 (continued)

Name	RDCS	IDCS	CBDS	HDDDCS
Iterative equations	$$\left.\begin{array}{l} x_1^n = F_1(\cdot) \\ x_2^n = F_2(\cdot) \\ \cdots \\ x_m^n = F_m(\cdot) \end{array}\right.$$	$$x_j^n = \begin{cases} F(x^{n-1})_j & \text{if } j = s^n \\ x_j^{n-1} & \text{if } j \neq s^n \end{cases}$$ $$j = N-1, N-2, \cdots, 0$$	$$x_j^n = \begin{cases} F(x^{n-1})_j & \text{if } s_j^n = 1 \\ x_j^{n-1} & \text{if } s_j^n = 0 \end{cases}$$ $$j = N-1, N-2, \cdots, 0$$	$$x_{1,j}^n = F_1(\cdot)s_j^n = \begin{cases} F_1(\cdot)_j & \text{if } s_j^n = 1 \\ x_{1,j}^{n-1} & \text{if } s_j^n = 0 \end{cases}$$ $$x_{2,j}^n = F_2(\cdot)u_j^n = \begin{cases} F_2(\cdot)_j & \text{if } u_j^n = 1 \\ x_{2,j}^{n-1} & \text{if } u_j^n = 0 \end{cases}$$ $$\cdots$$ $$x_{m,j}^n = F_m(\cdot)v_j^n = \begin{cases} F_m(\cdot)_j & \text{if } v_j^n = 1 \\ x_{m,j}^{n-1} & \text{if } v_j^n = 0 \end{cases}$$ $$j = P-1, P-2, \cdots, 0, -1, -2, \cdots, -Q$$

- RDCS present the finite integer part and infinite fractional part of a real number as

$$
\begin{cases}
x_1^n = x_{1,P-1}^n \cdots x_{1,0}^n . x_{1,-1}^n x_{1,-2}^n \cdots x_{1,-Q}^n \cdots, \\
x_2^n = x_{2,P-1}^n \cdots x_{2,0}^n . x_{2,-1}^n x_{2,-2}^n \cdots x_{2,-Q}^n \cdots, \\
\quad \vdots \\
x_m^n = x_{m,P-1}^n \cdots x_{m,0}^n . x_{m,-1}^n x_{m,-2}^n \cdots x_{m,-Q}^n \cdots.
\end{cases}
$$

But IDCS and CBDS process data on the integer domain as

$$
x^n = x_{N-1}^n x_{N-2}^n \cdots x_0^n.
$$

Furthermore, in HDDCS, the range of the input data is extended to the digital domain of the finite integer part and finite fractional part as

$$
\begin{cases}
x_1^n = x_{1,P-1}^n x_{1,P-2}^n \cdots x_{1,0}^n . x_{1,-1}^n x_{1,-2}^n \cdots x_{1,-Q}^n, \\
x_2^n = x_{2,P-1}^n x_{2,P-2}^n \cdots x_{2,0}^n . x_{2,-1}^n x_{2,-2}^n \cdots x_{2,-Q}^n, \\
\quad \vdots \\
x_m^n = x_{m,P-1}^n x_{m,P-2}^n \cdots x_{m,0}^n . x_{m,-1}^n x_{m,-2}^n \cdots x_{m,-Q}^n,
\end{cases}
$$

where $N = P + Q$.
- Both IDCS and CBDS deal with 1D chaotic systems. In contrast, HDDCS extends the system to any finite dimension.
- Compared to RDCS, IDCS, CBDS, and HDDCS need external inputs.
- The main features of discrete-time RDCS is that all the bits in x_{n-1} will be updated by the iterative equation F at each iteration operation. Likewise, all the bits in x_n are updated by the iterative equation F at each iteration operation. But only one bit of x_n in IDCS is updated by the iterative equation F at each iteration operation. CBDS uses multiple random bitwise operations instead of only one in IDCS. HDDCS are similar to CBDS but can work in higher dimensions.

5.1.4 Network Analysis of the State Space of HDDCS

Given a digital chaotic system, a state and its interval are mapped to another one. Considering each state or interval as a node (vertex), and the mapping relation as a directed edge (link), the state network of the chaotic system can be built up. As shown in [8, 9], the associated state network can demonstrate some dynamical properties of digital chaotic systems that cannot be observed by the previous analytic methods. For the directed graph of G_F in HDDCS, all the possible combinations of $(x_1, x_2, \ldots, x_m)$

are the nodes, and there is a directed edge from node $(\hat{x}_1, \hat{x}_2, \ldots, \hat{x}_m)$ to another node $(\tilde{x}_1, \tilde{x}_2, \ldots, \tilde{x}_m)$ if

$$(G_F((\hat{s}, \hat{u}, \ldots, \hat{v}), (\hat{x}_1, \hat{x}_2, \ldots, \hat{x}_m)))_{x_1, x_2, \ldots, x_m} = (\tilde{x}_1, \tilde{x}_2, \ldots, \tilde{x}_m).$$

We found that the state network of HDDCS must be strongly connected; namely every node is reachable from any other one.

In the following, we use 2D-DCS with $N = 2$ ($P = 2$, $Q = 0$) to illustrate the property on connectivity. For example, consider a 2D digital system uncontrolled by random sequences, as

Table 5.2 State transition for system (5.12). (©IEEE 2016, Reproduced with permission from [5])

(x^{n-1}, y^{n-1})	(x^n, y^n)	(x^{n-1}, y^{n-1})	(x^n, y^n)
(0, 0)	(3, 3)	(2, 0)	(1, 1)
(0, 1)	(3, 2)	(2, 1)	(1, 0)
(0, 2)	(3, 1)	(2, 2)	(1, 3)
(0, 3)	(3, 0)	(2, 3)	(1, 2)
(1, 0)	(2, 2)	(3, 0)	(0, 0)
(1, 1)	(2, 3)	(3, 1)	(0, 1)
(1, 2)	(2, 0)	(3, 2)	(0, 2)
(1, 3)	(2, 1)	(3, 3)	(0, 3)

Fig. 5.2 The disconnected network for Eq. (5.12). (©IEEE 2016, Reproduced with permission from [5])

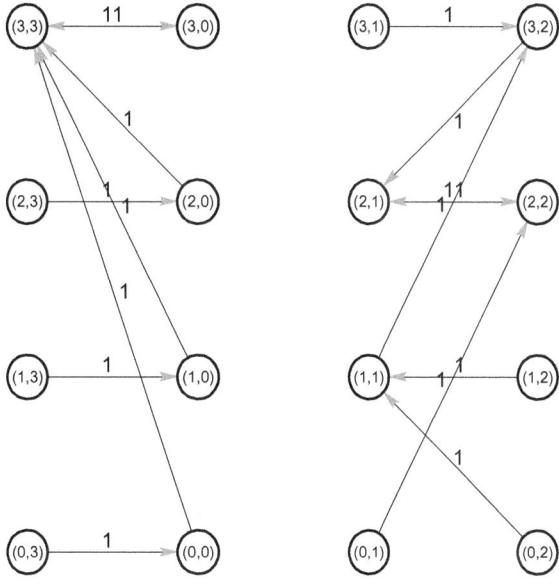

Table 5.3 State Transition of G_F for system (5.12) in HDDCS. (©IEEE 2016, Reproduced with permission from [5])

(x^{n-1}, y^{n-1}) \\ (s^n, u^n)	(0,0)	(0,1)	(0,2)	(0,3)	(1,0)	(1,1)	(1,2)	(1,3)	(2,0)	(2,1)	(2,2)	(2,3)	(3,0)	(3,1)	(3,2)	(3,3)
(0,0)	(0,0)	(0,1)	(0,2)	(0,3)	(1,0)	(1,1)	(1,2)	(1,3)	(2,0)	(2,1)	(2,2)	(2,3)	(3,0)	(3,1)	(3,2)	(3,3)
(0,1)	(0,1)	(0,0)	(0,3)	(0,2)	(1,1)	(1,0)	(1,3)	(1,2)	(2,1)	(2,0)	(2,3)	(2,2)	(3,1)	(3,0)	(3,3)	(3,2)
(0,2)	(0,2)	(0,3)	(0,0)	(0,1)	(1,2)	(1,3)	(1,0)	(1,1)	(2,2)	(2,3)	(2,0)	(2,1)	(3,2)	(3,3)	(3,0)	(3,1)
(0,3)	(0,3)	(0,2)	(0,1)	(0,0)	(1,3)	(1,2)	(1,1)	(1,0)	(2,3)	(2,2)	(2,1)	(2,0)	(3,3)	(3,2)	(3,1)	(3,0)
(1,0)	(1,0)	(1,1)	(1,2)	(1,3)	(0,0)	(0,1)	(0,2)	(0,3)	(3,0)	(3,1)	(3,2)	(3,3)	(2,0)	(2,1)	(2,2)	(2,3)
(1,1)	(1,1)	(1,0)	(1,3)	(1,2)	(0,1)	(0,0)	(0,3)	(0,2)	(3,1)	(3,0)	(3,3)	(3,2)	(2,1)	(2,0)	(2,3)	(2,2)
(1,2)	(1,2)	(1,3)	(1,0)	(1,1)	(0,2)	(0,3)	(0,0)	(0,1)	(3,2)	(3,3)	(3,0)	(3,1)	(2,2)	(2,3)	(2,0)	(2,1)
(1,3)	(1,3)	(1,2)	(1,1)	(1,0)	(0,3)	(0,2)	(0,1)	(0,0)	(3,3)	(3,2)	(3,1)	(3,0)	(2,3)	(2,2)	(2,1)	(2,0)
(2,0)	(2,0)	(2,1)	(2,2)	(2,3)	(3,0)	(3,1)	(3,2)	(3,3)	(0,0)	(0,1)	(0,2)	(0,3)	(1,0)	(1,1)	(1,2)	(1,3)
(2,1)	(2,1)	(2,0)	(2,3)	(2,2)	(3,1)	(3,0)	(3,3)	(3,2)	(0,1)	(0,0)	(0,3)	(0,2)	(1,1)	(1,0)	(1,3)	(1,2)
(2,2)	(2,2)	(2,3)	(2,0)	(2,1)	(3,2)	(3,3)	(3,0)	(3,1)	(0,2)	(0,3)	(0,0)	(0,1)	(1,2)	(1,3)	(1,0)	(1,1)
(2,3)	(2,3)	(2,2)	(2,1)	(2,0)	(3,3)	(3,2)	(3,1)	(3,0)	(0,3)	(0,2)	(0,1)	(0,0)	(1,3)	(1,2)	(1,1)	(1,0)
(3,0)	(3,0)	(3,1)	(3,2)	(3,3)	(2,0)	(2,1)	(2,2)	(2,3)	(1,0)	(1,1)	(1,2)	(1,3)	(0,0)	(0,1)	(0,2)	(0,3)
(3,1)	(3,1)	(3,0)	(3,3)	(3,2)	(2,1)	(2,0)	(2,3)	(2,2)	(1,1)	(1,0)	(1,3)	(1,2)	(0,1)	(0,0)	(0,3)	(0,2)
(3,2)	(3,2)	(3,3)	(3,0)	(3,1)	(2,2)	(2,3)	(2,0)	(2,1)	(1,2)	(1,3)	(1,0)	(1,1)	(0,2)	(0,3)	(0,0)	(0,1)
(3,3)	(3,3)	(3,2)	(3,1)	(3,0)	(2,3)	(2,2)	(2,1)	(2,0)	(1,3)	(1,2)	(1,1)	(1,0)	(0,3)	(0,2)	(0,1)	(0,0)

Fig. 5.3 The strongly connected network for G_F of system (5.12). (©IEEE 2016, Reproduced with permission from [5])

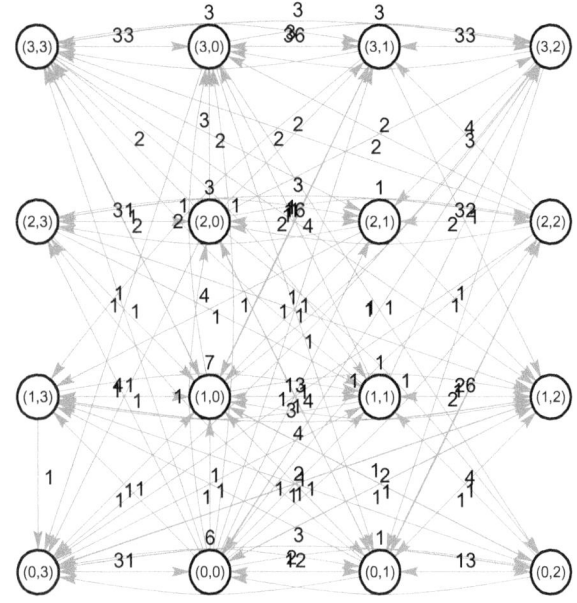

$$\begin{cases} x^n = F_1(x^{n-1}, y^{n-1}) = \overline{x^{n-1}}, \\ y^n = F_2(x^{n-1}, y^{n-1}) = \overline{x^{n-1}} \oplus y^{n-1}, \end{cases} \quad (5.12)$$

where $\oplus$ denotes the bitwise XOR.

The concrete state transition is shown in Table 5.2, and the corresponding state transition diagram is shown in Fig. 5.2.

According to Eqs. (5.10), (5.11), and (5.12), one can obtain the iterative form of $G_F(E)_{x,y}$ as

$$\begin{cases} x^n = x^{n-1} \cdot \overline{s^n} + (\overline{x^{n-1}} \cdot s^n), \\ y^n = y^{n-1} \cdot \overline{u^n} + ((x^{n-1} \oplus y^{n-1}) \cdot u^n), \end{cases} \quad (5.13)$$

where $s = s^1 s^2 s^3 \ldots$ and $u = u^1 u^2 u^3 \ldots$ are two random sequences generated by TRNG.

According to Eq. (5.13), the concrete state transition of $G_F(E)_{x,y}$ is shown in Table 5.3, and the corresponding state network is shown in Fig. 5.3, which is a strongly connected network. In the next section, we prove that HDDCS satisfies Devaney's definition of chaos if the state network of its G_F is strongly connected.

5.2 Chaotic Performance of HDDCS

In this section, we prove that the map of HDDCS satisfies Devaney's definition of chaos; that is, its periodic points are dense on its definitional domain and it is transitive [10].

5.2.1 Dense Periodic Points of HDDCS

To prove the existence of a dense set of periodic points and the transitivity property of the discrete dynamical system defined previously, the following lemma is first proven.

Lemma 5.1 *Let $w \in \{s, u, \cdots, v\}$, $w = w^1 w^2 w^3 \ldots w^n \ldots$ and $\hat{w} = \hat{w}^1 \hat{w}^2 \hat{w}^3 \ldots$ $\hat{w}^n \ldots$; the metric distance d satisfies that if $w^i = \hat{w}^i$ for $i = 1, 2, 3, \ldots n$, then*

$$d(w, \hat{w}) \leq \frac{1}{2^{Nn}},$$

where $w^k, \hat{w}^k \in [0, 2^P - 2^{-Q}]$ for $k \in \mathbb{Z}^+$.

Proof If $w^i = \hat{w}^i$ for $i = 1, 2, \ldots, n$, then

$$d(w, \hat{w}) = \sum_{i=1}^{n} \frac{|w^i - \hat{w}^i|}{2^{Ni}} + \sum_{i=n+1}^{\infty} \frac{|w^i - \hat{w}^i|}{2^{Ni}}$$

$$= \sum_{i=n+1}^{\infty} \frac{|w^i - \hat{w}^i|}{2^{Ni}}$$

$$\leq \sum_{i=n+1}^{\infty} \frac{2^P - 2^{-Q}}{2^{Ni}}$$

$$= \frac{2^P - 2^{-Q}}{2^N - 1} \cdot \frac{1}{2^{Nn}} \leq \frac{1}{2^{Nn}}.$$

Due to the definition of the proposed distance: for any $m \leq n$, if $w^m = \hat{w}^m$, then $d(w, \hat{w}) \leq \frac{1}{2^{Nn}}$.

The lemma can let us quickly determine whether the two sequences are close to each other. From intuitive observation, we can ensure two sequences are close to each other as long as they have a considerable number of consistent foregoing entries.

Theorem 5.1 *The periodic points of HDDCS are dense in the metric space $(\mathscr{E}, d)$.*

Proof For any given $\varepsilon \in (0, 1)$ and a periodic point $((\tilde{s}, \tilde{u}, \ldots, \tilde{v})(\tilde{x}_1, \tilde{x}_2, \ldots, \tilde{x}_m)) \in \mathscr{E}$, let us try to prove that the dense periodic points in $(\mathscr{E}, d)$ can always be found in the neighborhood of distance ε of any point $((\hat{s}, \hat{u}, \ldots, \hat{v}), (\hat{x}_1, \hat{x}_2, \ldots, \hat{x}_m)) \in \mathscr{E}$ as

$$d(((\tilde{s}, \tilde{u}, \ldots, \tilde{v}), (\tilde{x}_1, \tilde{x}_2, \ldots, \tilde{x}_m)), ((\hat{s}, \hat{u}, \ldots, \hat{v}), (\hat{x}_1, \hat{x}_2, \ldots, \hat{x}_m))) < \varepsilon. \quad (5.14)$$

Without loss of generality, we assume that the general form of

$$((\hat{s}, \hat{u}, \ldots, \hat{v}), (\hat{x}_1, \hat{x}_2, \ldots, \hat{x}_m)))$$

is

$$((\hat{s}, \hat{u}, \ldots, \hat{v}), (\hat{x}_1, \hat{x}_2, \ldots, \hat{x}_m)))$$
$$= (((\hat{s}^1 \hat{s}^2 \ldots \hat{s}^{k_0} \ldots \hat{s}^n \ldots), (\hat{u}^1 \hat{u}^2 \ldots \hat{u}^{k_0} \ldots \hat{u}^n \ldots),$$
$$\ldots, (\hat{v}^1 \hat{v}^2 \ldots \hat{v}^{k_0} \ldots \hat{v}^n \ldots)), (\hat{x}_1, \hat{x}_2, \ldots, \hat{x}_m)) \in \mathscr{E}.$$

Given $\varepsilon < 2^{-Q}$, if point $(\hat{x}_1, \hat{x}_2, \ldots, \hat{x}_m)$, and $(\tilde{x}_1, \tilde{x}_2, \ldots, \tilde{x}_m)$ do not coincide in the m-dimensional space, we can obtain $(\tilde{x}_1, \tilde{x}_2, \ldots, \tilde{x}_m) \neq (\hat{x}_1, \hat{x}_2, \ldots, \hat{x}_m)$ such that

$$d_x((\tilde{x}_1, \tilde{x}_2, \ldots, \tilde{x}_m), (\hat{x}_1, \hat{x}_2, \ldots, \hat{x}_m)) = \sqrt{(\tilde{x}_1 - \hat{x}_1)^2 + (\tilde{x}_2 - \hat{x}_2)^2 + \ldots + (\tilde{x}_m - \hat{x}_m)^2}$$
$$\geq 2^{-Q}.$$

Then, one has

$$d(((\hat{s}, \hat{u}, \ldots, \hat{v}), (\hat{x}_1, \hat{x}_2, \ldots, \hat{x}_m)), ((\tilde{s}, \tilde{u}, \ldots, \tilde{v}), (\tilde{x}_1, \tilde{x}_2, \ldots, \tilde{x}_m))) \geq \varepsilon.$$

Thus, to satisfy

$$d(((\hat{s}, \hat{u}, \ldots, \hat{v}), (\hat{x}_1, \hat{x}_2, \ldots, \hat{x}_m)), ((\tilde{s}, \tilde{u}, \ldots, \tilde{v}), (\tilde{x}_1, \tilde{x}_2, \ldots, \tilde{x}_m))) < \varepsilon,$$

we must first set $\tilde{x}_1 = \hat{x}_1, \tilde{x}_2 = \hat{x}_2, \ldots, \tilde{x}_m = \hat{x}_m$. In the case

$$d(((\hat{s}, \hat{u}, \ldots, \hat{v}), (\hat{x}_1, \hat{x}_2, \ldots, \hat{x}_m)), ((\tilde{s}, \tilde{u}, \ldots, \tilde{v}), (\tilde{x}_1, \tilde{x}_2, \ldots, \tilde{x}_m))) < \varepsilon,$$

we should consider proving that $((\tilde{s}, \tilde{u}, \ldots, \tilde{v}), (\tilde{x}_1, \tilde{x}_2, \ldots, \tilde{x}_m))$ is a periodic point. If the first k_0 elements of $\hat{s}$ and $\tilde{s}$ are the same, then $d_s(\hat{s}, \tilde{s}) < 2^{-Nk_0} < \varepsilon$. Referring to Lemma 5.1, similar results can be obtained as k_0 elements of $\hat{u}, \ldots, \hat{v}$, and $\tilde{u}, \ldots, \tilde{v}$ are the same and

$$d_u(\hat{u}, \tilde{u}) < 2^{-Nk_0} < \varepsilon, \ldots, d_v(\hat{v}, \tilde{v}) < 2^{-Nk_0} < \varepsilon.$$

Thus, $\forall \varepsilon < 1$, an integer k_0 satisfying the relation $d_s(\hat{s}, \tilde{s}) + d_u(\hat{u}, u) + \ldots + d_v(\hat{v}, \tilde{v}) < m \times 2^{-Nk_0} < \varepsilon$ can always be found. For instance, to make $m \times 2^{-Nk_0} < \varepsilon$ hold, the value of k_0 can be set as

$$k_0 = \lfloor (\log_2 m - \log_2 \varepsilon)/N \rfloor + 1.$$

After the k_0th iteration, one has

$$(G_F^{k_0}((\tilde{s}, \tilde{u}, \ldots, \tilde{v}), (\hat{x}_1, \hat{x}_2, \ldots, \hat{x}_m)))_{x_1, x_2, \ldots, x_m} = (\hat{x}_1, \hat{x}_2, \ldots, \hat{x}_m).$$

The above equation shows that HDDCS starts from $(\hat{x}_1, \hat{x}_2, \ldots, \hat{x}_m)$, then returns back to it after k_0 iterations. This means that a periodic point $(\hat{x}_1, \hat{x}_2, \ldots, \hat{x}_m)$ is found out:

$$((\tilde{s}, \tilde{u}, \ldots, \tilde{v}), (\hat{x}_1, \hat{x}_2, \ldots, \hat{x}_m))$$
$$= (((s^1 s^2 \ldots s^{k_0} s^1 s^2 \ldots s^{k_0} \ldots), (u^1 u^2 \ldots u^{k_0} u^1 u^2 \ldots u^{k_0} \ldots),$$
$$\ldots, (v^1 v^2 \ldots v^{k_0} v^1 v^2 \ldots v^{k_0} \ldots)), (\hat{x}_1, \hat{x}_2, \ldots, \hat{x}_m)),$$

which satisfies

$$G_F^{k_0}((\tilde{s}, \tilde{u}, \ldots, \tilde{v}), (\hat{x}_1, \hat{x}_2, \ldots, \hat{x}_m)) = ((\tilde{s}, \tilde{u}, \ldots, \tilde{v}), (\hat{x}_1, \hat{x}_2, \ldots, \hat{x}_m)),$$

and

$$d(((\hat{s}, \hat{u}, \ldots, \hat{v}), (\hat{x}_1, \hat{x}_2, \ldots, \hat{x}_m)), ((\tilde{s}, \tilde{u}, \ldots, \tilde{v}), (\tilde{x}_1, \tilde{x}_2, \ldots, \tilde{x}_m)))$$
$$= d_s(\hat{s}, \tilde{s}) + d_u(\hat{u}, \tilde{u}) + \ldots + d_v(\hat{v}, \tilde{v}) + d_x((\hat{x}_1, \hat{x}_2, \ldots, \hat{x}_m), (\hat{x}_1, \hat{x}_2, \ldots, \hat{x}_m))$$
$$= d_s(\hat{s}, \tilde{s}) + d_u(\hat{u}, \tilde{u}) + \ldots + d_v(\hat{v}, \tilde{v})$$
$$< \varepsilon.$$

After the k_0th iteration, one gets

$$(G_F^{k_0}((\tilde{s}, \tilde{u}, \ldots, \tilde{v}), (\hat{x}_1, \hat{x}_2, \ldots, \hat{x}_m)))_{x_1, x_2, \ldots, x_m} \neq (\hat{x}_1, \hat{x}_2, \ldots, \hat{x}_m)$$

and

$$(G_F^{k_0}((\tilde{s}, \tilde{u}, \ldots, \tilde{v}), (\hat{x}_1, \hat{x}_2, \ldots, \hat{x}_m)))_{x_1, x_2, \ldots, x_m} = (x_1', x_2', \ldots, x_m').$$

Because G_F is strongly connected, there is at least one path from the state $(x_1', x_2', \ldots, x_m')$ to the state $(\hat{x}_1, \hat{x}_2, \ldots, \hat{x}_m)$, after another iteration of i_0 times, where i_0 is equal to the number of edges in the connected path between $(x_1', x_2', \ldots, x_m')$ and $(\hat{x}_1, \hat{x}_2, \ldots, \hat{x}_m)$. By making the equation

$$(G_F^{k_0 + i_0}((\tilde{s}, \tilde{u}, \ldots, \tilde{v}), (\hat{x}_1, \hat{x}_2, \ldots, \hat{x}_m)))_{x_1, x_2, \ldots, x_m} = (\hat{x}_1, \hat{x}_2, \ldots, \hat{x}_m)$$

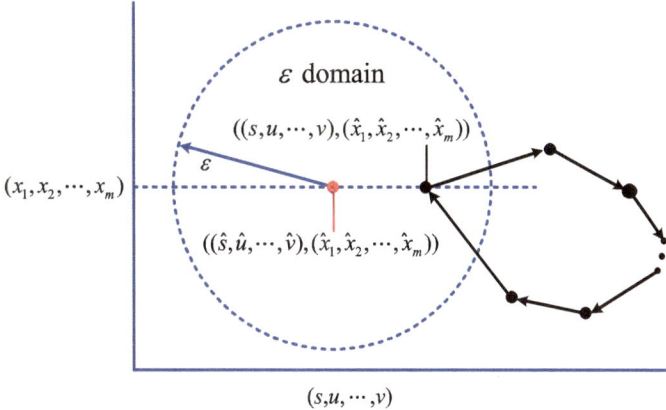

Fig. 5.4 The diagram of the periodic points of HDDCS. (©IEEE 2016, Reproduced with permission from [5])

hold, a periodic point is found by checking

$$((\tilde{s}, \tilde{u}, \ldots, \tilde{v}), (\hat{x}_1, \hat{x}_2, \ldots, \hat{x}_m)) = (((s^1 s^2 \ldots s^{k_0} s^{k_0+1} s^{k_0+2} \ldots s^{k_0+i_0} s^1 s^2 \ldots s^{k_0}$$
$$s^{k_0+1} s^{k_0+2} \ldots s^{k_0+i_0} \ldots), (u^1 u^2 \ldots u^{k_0} u^{k_0+1}$$
$$u^{k_0+2} \ldots u^{k_0+i_0} u^1 u^2 \ldots u^{k_0} u^{k_0+1} u^{k_0+2} \ldots$$
$$u^{k_0+i_0} \ldots), \ldots, (v^1 v^2 \ldots v^{k_0} v^{k_0+1} v^{k_0+2} \ldots$$
$$v^{k_0+i_0} v^1 v^2 \ldots v^{k_0} v^{k_0+1} v^{k_0+2} \ldots v^{k_0+i_0} \ldots))$$
$$(\hat{x}_1, \hat{x}_2, \ldots, \hat{x}_m)),$$

which satisfies

$$d(((\hat{s}, \hat{u}, \ldots, \hat{v}), (\hat{x}_1, \hat{x}_2, \ldots, \hat{x}_m))), ((\tilde{s}, \tilde{u}, \ldots, \tilde{v}), (\tilde{x}_1, \tilde{x}_2, \ldots, \tilde{x}_m)))$$
$$= d_s(\hat{s}, \tilde{s}) + d_u(\hat{u}, \tilde{u}) + \ldots + d_v(\hat{v}, \tilde{v}) + d_x((\hat{x}_1, \hat{x}_2, \ldots, \hat{x}_m), (\tilde{x}_1, \tilde{x}_2, \ldots, \tilde{x}_m))$$
$$= d_s(\hat{s}, \tilde{s}) + d_u(\hat{u}, \tilde{u}) + \ldots + d_v(\hat{v}, \tilde{v})$$
$$< \varepsilon.$$

In summary, the periodic points of G_F are dense in metric space $(\mathcal{E}, d)$, as shown in Fig. 5.4.

5.2.2 Transitive Property of HDDCS

Theorem 5.2 *Function G_F is topological transitive in the metric space $(\mathcal{E}, d)$.*

Proof Let us recall that the so-called topological transitivity of function G_F in metric space $(\mathcal{E}, d)$ means that there always exists $n_0 > 0$ satisfying $G_F^{n_0}(U') \cap U'' \neq \varnothing$ for any nonempty open sets U' and U''. Let $((s', u', \ldots, v'), (x_1', x_2', \ldots, x_m')) \in U' \subseteq \mathscr{E}''$ denote the center of U', which can be represented as

$$
\begin{aligned}
((s', u', \ldots, v'), (x_1', x_2', \ldots, x_m')) &= (((s'^1 s'^2 \ldots s'^{n_0} \ldots s'^n \ldots), \\
&\quad (u'^1 u'^2 \ldots u'^{n_0} \ldots u'^n \ldots), \ldots, \\
&\quad (v'^1 v'^2 \ldots v'^{n_0} \ldots v'^n \ldots)), \\
&\quad (x_1', x_2', \ldots, x_m').
\end{aligned}
$$

The center of U'', $((s'', u'', \ldots, v''), (x_1'', x_2'', \ldots, x_m''))$ can also be presented as

$$
\begin{aligned}
((s'', u'', \ldots, v''), (x_1'', x_2'', \ldots, x_m'')) &= (((s''^1 s''^2 \ldots s''^{n_0} \ldots s''^n \ldots), \\
&\quad u''^1 u''^2 \ldots u''^{n_0} \ldots u''^n \ldots), \ldots, \\
&\quad (v''^1 v''^2 \ldots v''^{n_0} \ldots v''^n \ldots)), \\
&\quad x_1'', x_2'', \ldots, x_m''),
\end{aligned}
$$

and a point in U' is denoted as $((\tilde{s}, \tilde{u}, \ldots, \tilde{v}), (\tilde{x}_1, \tilde{x}_2, \ldots, \tilde{x}_m)) \in U' \subseteq \mathscr{E}$.

If the sphere radius of U', r', is less than 2^{-Q} and points $(x_1', x_2', \ldots, x_m')$ and $(\tilde{x}_1, \tilde{x}_2, \ldots, \tilde{x}_m)$ do not coincide in the m-dimensional space, one has

$$
((\tilde{s}, \tilde{u}, \ldots, \tilde{v}), (\tilde{x}_1, \tilde{x}_2, \ldots, \tilde{x}_m) \notin U'
$$

as

$$
\begin{aligned}
d_x((\tilde{x}_1, \tilde{x}_2, \ldots, \tilde{x}_m), (x_1', x_2', \ldots, x_m')) &= \sqrt{(\tilde{x}_1 - x_1')^2 + (\tilde{x}_2 - x_2')^2 + \ldots + (\tilde{x}_m - x_m')^2} \\
&\geq 2^{-Q} > r',
\end{aligned}
$$

and $(\tilde{x}_1, \tilde{x}_2, \ldots, \tilde{x}_m) \neq (x_1', x_2', \ldots, x_m')$. Thus to satisfy

$$
((\tilde{s}, \tilde{u}, \ldots, \tilde{v}), (\tilde{x}_1, \tilde{x}_2, \ldots, \tilde{x}_m)) \in U',
$$

we must set

$$
(\tilde{x}_1, \tilde{x}_2, \ldots, \tilde{x}_m) = (x_1', x_2', \ldots, x_m')
$$

to obtain

$$
d_x((\tilde{x}_1, \tilde{x}_2, \ldots, \tilde{x}_m), (x_1', x_2', \ldots, x_m')) = 0.
$$

If the first k_0 elements of s' and $\tilde{s}$ are the same, then $d_s(s', \tilde{s}) < 2^{-Nk_0}$. From Lemma 5.1, similar results can be obtained when the first k_0 elements of $u', \ldots, v'$ and $\tilde{u}, \ldots, \tilde{v}$ are the same and

$$
d_u(u', \tilde{u}) < 2^{-Nk_0} < \varepsilon, \ldots, d_v(v', \tilde{v}) < 2^{-Nk_0} < \varepsilon.
$$

Therefore, $\forall\, r' < 1$, an integer k_0 satisfying $d_s(s', \tilde{s}) + d_u(u', \tilde{u}) + \ldots + d_v(v', \tilde{v}) < m \times 2^{-Nk_0} < r'$ can always be found. For instance, to satisfy $m \times 2^{-Nk_0} < r'$, the value of k_0 can be set as

$$k_0 = \lfloor (\log_2 m - \log_2 r')/N \rfloor + 1.$$

If after the k_0th iteration, equality

$$(G_F^{k_0}((\tilde{s}, \tilde{u}, \ldots, \tilde{v}), (\tilde{x}_1, \tilde{x}_2, \ldots, \tilde{x}_m)))_{x_1, x_2, \ldots, x_m} = (x_1'', x_2'', \ldots, x_m'')$$

exists, then $n_0 = k_0$ is found, and

$$
\begin{aligned}
((\tilde{s}, \tilde{u}, \ldots, \tilde{v}), (\tilde{x}_1, \tilde{x}_2, \ldots, \tilde{x}_m)) = (((s'^1 s'^2 \ldots s'^{n_0} s''^1 s''^2 \ldots s''^n \ldots), (u'^1 u'^2 \ldots u'^{n_0} \\
u''^1 u''^2 \ldots u''^n \ldots), \ldots, (v'^1 v'^2 \ldots v'^{n_0} v''^1 v''^2 \ldots \\
v''^n \ldots)), (x_1', x_2', \ldots, x_m')) \in U',
\end{aligned}
$$

which satisfies

$$
\begin{aligned}
G_F^{n_0}((\tilde{s}, \tilde{u}, \ldots, \tilde{v}), (\tilde{x}_1, \tilde{x}_2, \ldots, \tilde{x}_m)) &= ((s'', u'', \ldots, v''), (x_1'', x_2'', \ldots, x_m'')) \\
&\in G_F^{n_0}(U') \cap U''.
\end{aligned}
$$

Thus

$$G_F^{n_0}(U') \cap U'' \neq \varnothing$$

holds.

If after the k_0th iteration, inequality

$$(G_F^{k_0}((\tilde{s}, \tilde{u}, \ldots, \tilde{v}), (\tilde{x}_1, \tilde{x}_2, \ldots, \tilde{x}_m)))_{x_1, x_2, \ldots, x_m} \neq (x_1'', x_2'', \ldots, x_m''),$$

holds, set $(G_F^{k_0}((\tilde{s}, \tilde{u}, \ldots, \tilde{v}), (\tilde{x}_1, \tilde{x}_2, \ldots, \tilde{x}_m)))_{x_1, x_2, \ldots, x_m} = (x_1''', x_2''', \ldots, x_m''')$. Because G_F is strongly connected, there is at least one path from the state $(x_1''', x_2''', \ldots, x_m''')$ to the state $(x_1'', x_2'', \ldots, x_m'')$ after another i_0 iterations. Because

$$G_F^{i_0}((s''', u''', \ldots, v'''), (x_1''', x_2''', \ldots, x_m''')) = ((s'', u'', \ldots, v''), (x_1'', x_2'', \ldots, x_m''))$$

holds, $n_0 = k_0 + i_0$ is found, and

$$
\begin{aligned}
((\tilde{s}, \tilde{u}, \ldots, \tilde{v}), (\tilde{x}_1, \tilde{x}_2, \ldots, \tilde{x}_m)) = (((s'^1 s'^2 \ldots s'^{k_0} s^{k_0+1} s^{k_0+2} \ldots s^{k_0+i_0} s'''^1 s''^2 \ldots \\
s'''^n \ldots), (u'^1 u'^2 \ldots u'^{k_0} u^{k_0+1} u^{k_0+2} \ldots u^{k_0+i_0} \\
u''^1 u''^2 \ldots u''^n \ldots), \ldots, (v'^1 v'^2 \ldots v'^{k_0} v^{k_0+1} \\
v^{k_0+2} \ldots v^{k_0+i_0} v''^1 v''^2 \ldots v''^n \ldots)), \\
(x_1', x_2', \ldots, x_m')) \\
\in U',
\end{aligned}
$$

which satisfies

$$G_F^{n_0}((\tilde{s}, \tilde{u}, \ldots, \tilde{v}), (\tilde{x}_1, \tilde{x}_2, \ldots, \tilde{x}_m)) = ((s'', u'', \ldots, v''), (x_1'', x_2'', \ldots, x_m''))$$
$$\in G_F^{n_0}(U') \cap U''.$$

Therefore one has

$$G_F^{n_0}(U') \cap U'' \neq \emptyset.$$

In summary, G_F is transitive in the metric space $(\mathscr{E}, d)$.

If a dynamical system on a metric space is transitive and has dense periodic points, then it has sensitive dependence on initial conditions [10]. In other words, if the state network of HDDCS is strongly connected, we can prove that it is chaotic in the sense of Devaney's definition of chaos.

5.3 Lyapunov Exponents of a Class of HDDCS

In this section, the Lyapunov exponents of HDDCS with $N = P$ ($Q = 0$) are estimated.

5.3.1 General Expression of Equivalent Decimal for G_F

Set the binary form of an m-dimensional array of N-bit integers as

$$\begin{cases} x_1 = x_{1,N-1} x_{1,N-2} \ldots x_{1,0}, \\ x_2 = x_{2,N-1} x_{2,N-2} \ldots x_{2,0}, \\ \quad \vdots \\ x_m = x_{m,N-1} x_{m,N-2} \ldots x_{m,0}, \end{cases}$$

where $x_{i,j} \in \{0, 1\}$ $(i = 1, 2, \ldots, m; j = N - 1, N - 2, \ldots, 0)$.

The general form of the corresponding decimal integer is obtained as

$$\begin{cases} X_1 = \sum_{k=0}^{N-1} (x_{1,k} \cdot 2^k). \\ X_2 = \sum_{k=0}^{N-1} (x_{2,k} \cdot 2^k). \\ \quad \vdots \\ X_m = \sum_{k=0}^{N-1} (x_{m,k} \cdot 2^k). \end{cases} \tag{5.15}$$

The random number of m sequences is expressed in corresponding decimal fraction as

$$\begin{cases} S = \sum_{k=1}^{+\infty}(s^k \cdot 2^{-Nk}), \\ U = \sum_{k=1}^{+\infty}(u^k \cdot 2^{-Nk}), \\ \quad \vdots \\ V = \sum_{k=1}^{+\infty}(v^k \cdot 2^{-Nk}), \end{cases} \qquad (5.16)$$

where $2^{-Nk}(k = 1, 2, \ldots)$ denote weights.

According to Eqs. (5.15) and (5.16) added together, the general form of the corresponding decimal number is obtained as

$$\begin{cases} y_1 = \sum_{k=0}^{N-1}(x_{1,k} \cdot 2^k) + \sum_{k=1}^{+\infty}(s^k \cdot 2^{-Nk}), \\ y_2 = \sum_{k=0}^{N-1}(x_{2,k} \cdot 2^k) + \sum_{k=1}^{+\infty}(u^k \cdot 2^{-Nk}), \\ \quad \vdots \\ y_m = \sum_{k=0}^{N-1}(x_{m,k} \cdot 2^k) + \sum_{k=1}^{+\infty}(v^k \cdot 2^{-Nk}). \end{cases}$$

According to the chaos generation strategy controlled by random sequences, the general form of an m-dimensional digital discrete-time iterative equation can be presented as

$$\begin{cases} g_1(X) = (x_1 \cdot \overline{s^1}) + (F_1(\cdot) \cdot s^1), \\ g_2(X) = (x_2 \cdot \overline{u^1}) + (F_2(\cdot) \cdot u^1), \\ \quad \vdots \\ g_m(X) = (x_m \cdot \overline{v^1}) + (F_m(\cdot) \cdot v^1), \end{cases} \qquad (5.17)$$

where $X = (X_1, X_2, \ldots, X_m)$ and

$$\begin{cases} F_1(\cdot) \triangleq F_1(x_1, x_2, \ldots, x_m), \\ F_2(\cdot) \triangleq F_2(x_1, x_2, \ldots, x_m), \\ \quad \vdots \\ F_m(\cdot) \triangleq F_m(x_1, x_2, \ldots, x_m). \end{cases}$$

There are separate shifts of one value in each one-sided infinite sequence ($s = s^1 s^2 \ldots s^n \ldots$, $u = u^1 u^2 \ldots u^n \ldots$, $\ldots$, $v = v^1 v^2 \ldots v^n \ldots$); the first value turns into $s^2, u^2, \ldots, v^2$ individually, and the corresponding weight is 2^{-N}. Therefore the general form of the corresponding decimal fraction after shifting one value in every one-sided infinite sequence is

$$\begin{cases} g_1(S) = 2^N \sum_{k=2}^{+\infty}(s^k \cdot 2^{-Nk}), \\ g_2(U) = 2^N \sum_{k=2}^{+\infty}(u^k \cdot 2^{-Nk}), \\ \quad \vdots \\ g_m(V) = 2^N \sum_{k=2}^{+\infty}(v^k \cdot 2^{-Nk}). \end{cases} \tag{5.18}$$

Adding Eqs. (5.17) and (5.18) together, one can obtain the general form of the corresponding decimal number,

$$\begin{cases} g_1(y_1, y_2, \ldots, y_m) = (x_1 \cdot \overline{s^1}) + (F_1(\cdot) \cdot s^1) + 2^N \sum_{k=2}^{+\infty}(s^k \cdot 2^{-Nk}), \\ g_2(y_1, y_2, \ldots, y_m) = (x_2 \cdot \overline{u^1}) + (F_2(\cdot) \cdot u^1) + 2^N \sum_{k=2}^{+\infty}(u^k \cdot 2^{-Nk}), \\ \quad \vdots \\ g_m(y_1, y_2, \ldots, y_m) = (x_m \cdot \overline{v^1}) + (F_m(\cdot) \cdot v^1) + 2^N \sum_{k=2}^{+\infty}(v^k \cdot 2^{-Nk}), \end{cases}$$

after randomly updating multiple random bits and shifting one value in every one-sided infinite sequence.

5.3.2 Mathematical Expression for $\frac{\partial g_i(y_1, y_2, \ldots, y_m)}{\partial y_j}$

In the interval $[\frac{n}{2^N}, \frac{n+1}{2^N})$ $(n \in [0, 2^{2N} - 1])$, the part of the decimal integer is not changed where $\Delta X_1 = \Delta X_2 = \cdots = \Delta X_m = 0$. Furthermore, the first decimals of the m sequences are the same; thus

$$\begin{cases} \Delta y_1 = \Delta X_1 + \Delta S = \Delta S = \sum_{k=2}^{+\infty}(\Delta s^k \cdot 2^{-Nk}), \\ \Delta y_2 = \Delta X_2 + \Delta U = \Delta U = \sum_{k=2}^{+\infty}(\Delta u^k \cdot 2^{-Nk}), \\ \quad \vdots \\ \Delta y_m = \Delta X_m + \Delta U = \Delta U = \sum_{k=2}^{+\infty}(\Delta v^k \cdot 2^{-Nk}). \end{cases}$$

From the definition of the partial derivative, one has

$$\begin{aligned} \frac{\partial g_1(y_1, y_2, \ldots, y_m)}{\partial y_1} &= \lim_{\Delta y_1 \to 0} \frac{g_1(y_1 + \Delta y_1, y_2, \cdots, y_m) - g_1(y_1, y_2, \cdots, y_m)}{\Delta y_1} \\ &= \lim_{\Delta S \to 0} \left(\frac{g_1(X) + 2^N \sum_{k=2}^{+\infty}((s^k + \Delta s^k) \cdot 2^{-Nk})}{\sum_{k=2}^{+\infty}(\Delta s^k) \cdot 2^{-Nk}} \right. \\ &\quad \left. - \frac{g_1(X) + 2^N \sum_{k=2}^{+\infty}(s^k \cdot 2^{-Nk})}{\sum_{k=2}^{+\infty}(\Delta s^k) \cdot 2^{-Nk}} \right) \\ &= \lim_{\Delta S \to 0} \frac{2^N \sum_{k=2}^{+\infty}(\Delta s^k \cdot 2^{-Nk})}{\sum_{k=2}^{+\infty}(\Delta s^k \cdot 2^{-Nk})} \\ &= 2^N. \end{aligned}$$

Similarly, one can get

$$\frac{\partial g_2(y_1, y_2, \cdots, y_m)}{\partial y_2} = \frac{\partial g_3(y_1, y_2, \cdots, y_m)}{\partial y_3}$$

$$= \cdots$$

$$= \frac{\partial g_m(y_1, y_2, \cdots, y_m)}{\partial y_m}$$

$$= 2^N.$$

On the other side, one has

$$\begin{cases} \dfrac{\partial g_1(y_1, y_2, \cdots, y_m)}{\partial y_2} = \dfrac{\partial g_1(y_1, y_2, \cdots, y_m)}{\partial y_3} = \cdots = \dfrac{\partial g_1(y_1, y_2, \cdots, y_m)}{\partial y_m} = 0, \\[2mm] \dfrac{\partial g_2(y_1, y_2, \cdots, y_m)}{\partial y_1} = \dfrac{\partial g_2(y_1, y_2, \cdots, y_m)}{\partial y_3} = \cdots = \dfrac{\partial g_2(y_1, y_2, \cdots, y_m)}{\partial y_m} = 0, \\[2mm] \qquad\qquad\qquad\qquad \vdots \\[2mm] \dfrac{\partial g_m(y_1, y_2, \cdots, y_m)}{\partial y_1} = \dfrac{\partial g_m(y_1, y_2, \cdots, y_m)}{\partial y_2} = \cdots = \dfrac{\partial g_m(y_1, y_2, \cdots, y_m)}{\partial y_{m-1}} = 0. \end{cases}$$

Thus the corresponding Jacobian matrix, diagonal matrix $\mathrm{diag}(2^N, 2^N, \cdots, 2^N)$, is obtained.

5.3.3 Estimating the Lyapunov Exponents

Let $\mu_k(\Phi_n^T \cdot \Phi_n)$ denote the kth characteristic value of matrix $(\Phi_n^T \cdot \Phi_n)$; the Lyapunov exponent of the specific HDDCS can be estimated as

$$\begin{aligned} \lambda(y_k) &= \lim_{n \to +\infty} \tfrac{1}{2n} \ln |\mu_k(\Phi_n^T \cdot \Phi_n)| \\ &= \lim_{n \to +\infty} \tfrac{1}{2n} \ln((2^N)^{2n}) \qquad\qquad (5.19) \\ &= N \ln 2, \end{aligned}$$

where $\Phi_n = J^n$ and $k = 1, \ldots, m$ [11].

Some concrete examples are provided to illustrate the parameters in the following Eq. (5.19).

1. Set $N = 2, m = 2$ and

$$\begin{cases} g_1(y_1, y_2) = (x_1 \cdot \overline{s^1}) + (\overline{x_1} \cdot s^1) + 4\sum_{k=2}^{+\infty}(4^{-k}s^k), \\ g_2(y_1, y_2) = (x_2 \cdot \overline{u^1}) + ((\overline{x_1} \oplus x_2) \cdot u^1) + 4\sum_{k=2}^{+\infty}(4^{-k}u^k), \end{cases} \qquad (5.20)$$

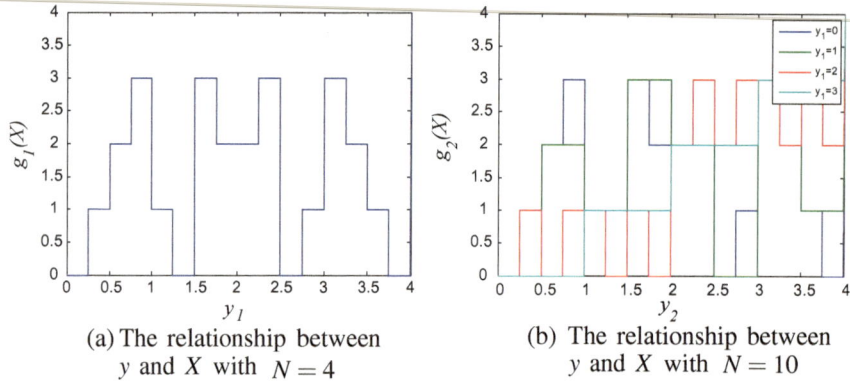

(a) The relationship between
y and X with N = 4

(b) The relationship between
y and X with N = 10

Fig. 5.5 The relationship between y and X

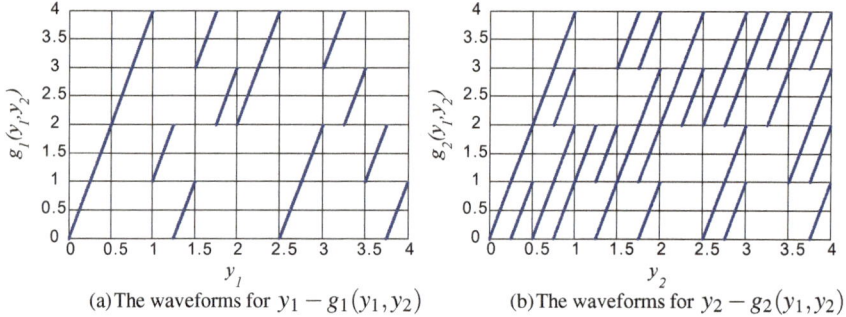

(a) The waveforms for $y_1 - g_1(y_1, y_2)$

(b) The waveforms for $y_2 - g_2(y_1, y_2)$

Fig. 5.6 The waveform

where

$$\begin{cases} g_1(X) = (x_1 \cdot \overline{s^1}) + (\overline{x_1} \cdot s^1), \\ g_2(X) = (x_2 \cdot \overline{u^1}) + ((\overline{x_1} \oplus x_2) \cdot u^1), \end{cases}$$

The relationship between y_1 and $g_1(X)$ is shown in Fig. 5.5a, therefore any two points in the interval $[\frac{n}{4}, \frac{n+1}{4})(n \in \{0, 1, \ldots, 2^4\})$ have the same integer part. Similarly, the relationship between y_2 and $g_2(X)$ is shown in Fig. 5.5b. In the general case N, any two points in the interval $[\frac{n}{2^N}, \frac{n+1}{2^N})(n \in \{0, 1, \ldots, 2^{2N} - 1\})$ have the same integer part at each dimension.

2. According to Eq. (5.20), the waveforms for $y_1 - g_1(y_1, y_2)$ and $y_2 - g_2(y_1, y_2)$ are shown in Fig. 5.6a, b. Obviously, except for the break point, the slope of the remaining part is 4, then $\frac{\partial g(y_1, y_2, \ldots, y_m)}{\partial y_k} = 4$ is established.

3. If the system is conservative (i.e., there is no dissipation), the Lyapunov exponents must be zero. If the system is dissipative, the Lyapunov exponent is negative. And

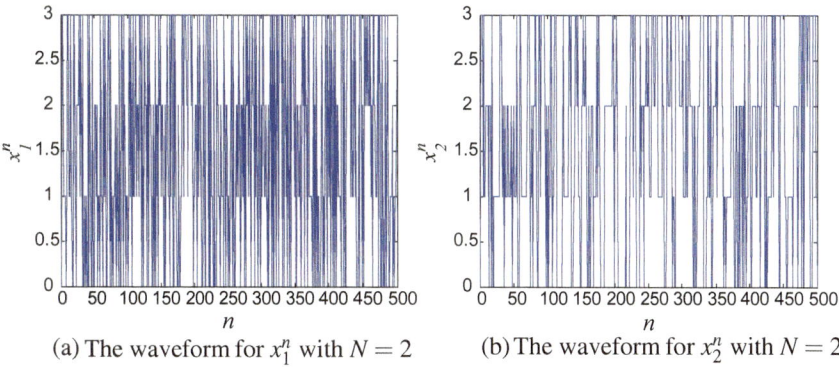

(a) The waveform for x_1^n with $N = 2$ (b) The waveform for x_2^n with $N = 2$

Fig. 5.7 The waveform for $N = 2$

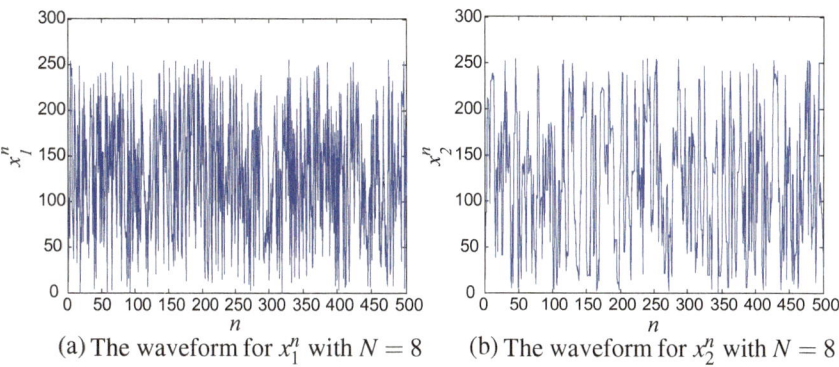

(a) The waveform for x_1^n with $N = 8$ (b) The waveform for x_2^n with $N = 8$

Fig. 5.8 The waveform for $N = 8$

a positive Lyapunov exponent is usually taken as an indication that the system is chaotic.

According to Eq. (5.19), a dynamical system with $N > 1$, all positive Lyapunov exponents $\lambda(y_k) > 0 (k = 1, 2, \ldots, m)$, and global bounded movement $|y_k^n| \leq 2^N (k = 1, 2, \ldots, m; n = 0, 1, 2, \ldots)$, displays chaotic behavior. For example, set $N = 2$ and Lyapunov exponents $\lambda(y_1) > 0$, $\lambda(y_2) > 0$ and $y_1^n \leq 3$, $y_2^n \leq 3$; then the simulated chaotic waveforms are shown in Fig. 5.7.

4. According to Eq. (5.19), the chaotic property is stronger as the Lyapunov exponent increases with the number N. Therefore, a larger N for the integer domain chaotic system is usually recommended in practical applications. The simulated chaotic waveform with $N = 8$ is shown in Fig. 5.8 for comparison.

5.4 FPGA-Based Real-Time Application of 3D-DCS

In this section, we first give the design and implementation of a FPGA (field programmable gate array)-based generator for 3D-DCS. Then, an RGB (red, green, blue) color image encryption method based on 3D-DCS is presented. We split the RGB color image into its three R, G, B components, and use the 3D-DCS to scramble the pixel values of the three components. Finally, a systematic methodology for FPGA platform-based implementation of the above method is proposed.

5.4.1 Design of 3D-DCS in FPGA

To use the power of FPGA, the computation needs to be divided into several independent blocks of threads that can be executed simultaneously. The performance on FPGA is directly related to the number of threads and that of logistical elements used during processing; its performance decreases when the number of branching instructions (moves such as while, if, etc.) increases. Following these rules, it is possible to build a Verilog-HDL program of the 3D-DCS algorithm [12].

A concrete example is provided to illustrate the 3D-DCS algorithm. Here, we use 3D-DCS with $N = 32$ ($P = 32$, $Q = 0$). For example, a 3D digital system controlled by random sequences is considered:

$$\begin{cases} x^n = \overline{x^{n-1}} \oplus (1 << (\mathrm{mod}\,(z^{n-1}, 32))), \\ y^n = \overline{y^{n-1}} \oplus (1 << (\mathrm{mod}\,(x^{n-1}, 32))), \\ z^n = \overline{z^{n-1}} \oplus (1 << (\mathrm{mod}\,(y^{n-1}, 32))). \end{cases} \tag{5.21}$$

where $\mathrm{mod}\,(m, n) = m - n \cdot \lfloor \frac{m}{n} \rfloor$, $\lfloor \frac{m}{n} \rfloor$ gives the largest integer less than or equal to $\frac{m}{n}$, and $<<$ represents the left bit shift. According to Eqs. (5.10), (5.11), and (5.21), one can obtain the iterative form of $G_F(E)_{x,y,z}$ as

$$\begin{cases} x^n = x^{n-1} \cdot \overline{s^n} + ((\overline{x^{n-1}} \oplus (1 << (\mathrm{mod}\,(z^{n-1}, 32)))) \cdot s^n), \\ y^n = y^{n-1} \cdot \overline{u^n} + ((\overline{y^{n-1}} \oplus (1 << (\mathrm{mod}\,(x^{n-1}, 32)))) \cdot u^n), \\ z^n = z^{n-1} \cdot \overline{v^n} + ((\overline{z^{n-1}} \oplus (1 << (\mathrm{mod}\,(y^{n-1}, 32)))) \cdot v^n). \end{cases} \tag{5.22}$$

where $s = s^1 s^2 s^3 \ldots$, $u = u^1 u^2 u^3 \ldots$, and $v = v^1 v^2 v^3 \ldots$ are three random sequences. This 3D-DCS may utilize any reasonable random sequence as s, u, v. For demonstration purposes, ISAAC (indirection, shift, accumulate, add, and count) in [13] is adopted here. It was found that the three channels of the outputs of 3D-DCS all can pass the NIST randomness test suite given in [14]. Meanwhile, the correlations among the three channels of the outputs and their autocorrelation strengths are all very low.

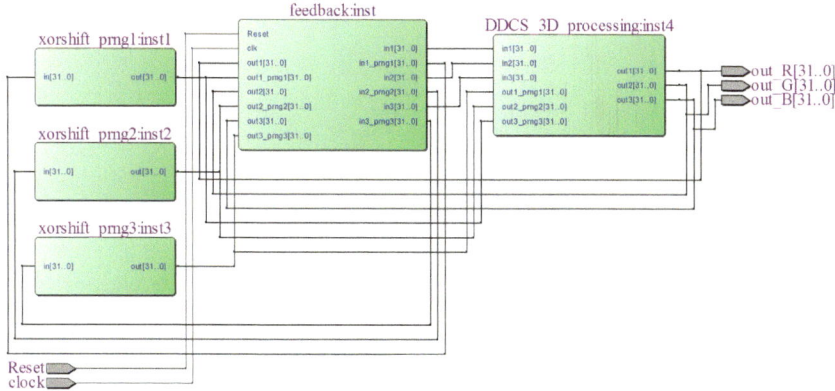

Fig. 5.9 Block Diagram of 3D-DCS in FPGA. (©IEEE 2016, Reproduced with permission from [5])

Figure 5.9 depicts the circuit structure of 3D-DCS. First of all, according to [4], three different types of oscillator ring TRNGs named *Oscillator_Rings1:inst1*, *Oscillator_Rings2:inst2*, and *Oscillator_Rings3:inst3* blocks shown in Fig. 5.9, are applied as the external control inputs s, u, and v. Then, the block *3D-DCS_processing:inst4* is constructed by using Eq. (5.22). Finally, the states of oscillator rings TRNGs and *3D-DCS_processing:inst4* are updated in the *feedback* block controlled by the clock signal, and it executes in clock positive edge.

The input clock frequency decides the speed of the 3D-DCS processing. In our experiments, the input clock is set at 50 MHz. Then, ModelSim Altera is used to obtain $x^n = out1$, $y^n = out2$, and $z^n = out3$ (Fig. 5.10), which are used for RGB image encryption and decryption processing.

5.4.2 Design of the FPGA-Based Hardware System for Image Encryption and Decryption

Figure 5.11 shows the block diagram for a FPGA-based hardware system of image encryption by 3D-DCS, where the function of the FPGA hardware part is implemented by the Verilog-HDL program.

As shown in Fig. 5.11, the hardware system at the transmitter side consists of five parts: picture RAM, 3D-DCS, VGA display controller, monitor 1, and the RGB encryption module. The picture is encrypted at the transmitter side and then transmitted through the public channel to the receiver. The receiver, for its part, contains four components: monitor 2, VGA display controller, RGB decryption module, and 3D-DCS.

The corresponding hardware implementation platform is shown in Fig. 5.12; the same two models of the Altera DE2 FPGA development board were used. The work-

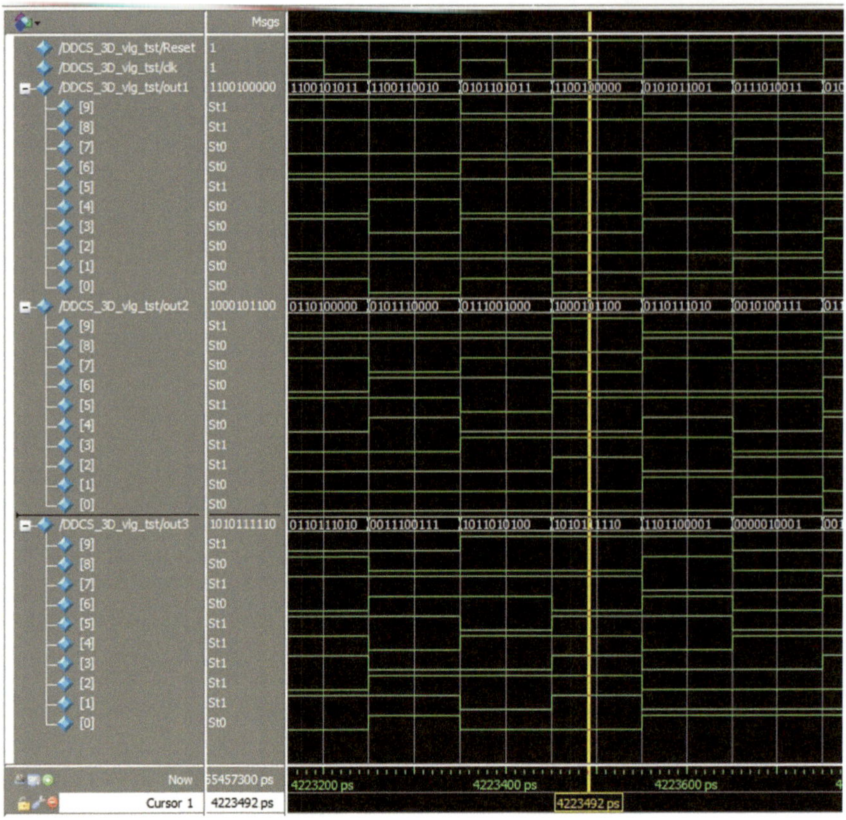

Fig. 5.10 ModelSim Altera simulation of 3D-DDCS in FPGA processing

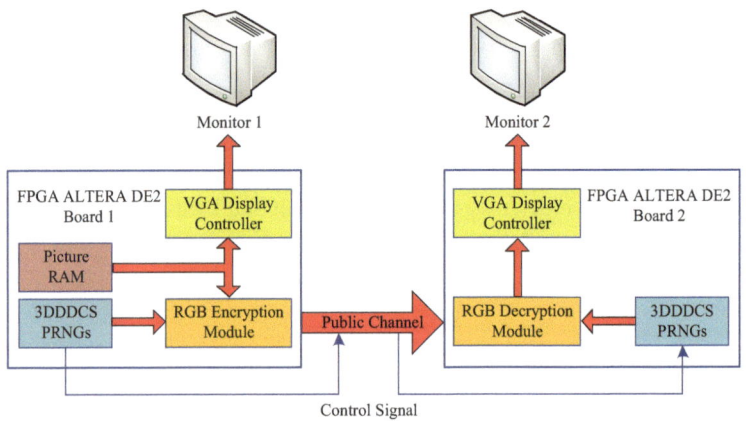

Fig. 5.11 Block diagram of FPGA-based application on image encryption. (©IEEE 2016, Reproduced with permission from [5])

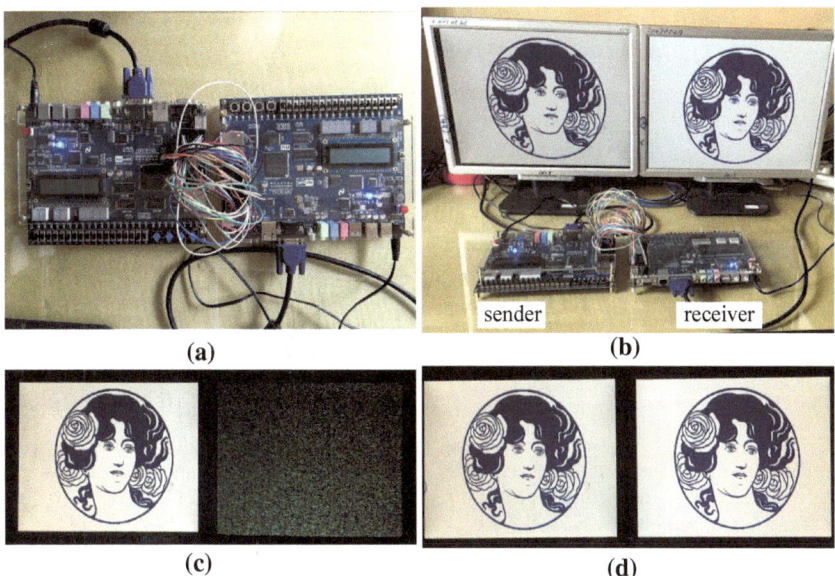

Fig. 5.12 FPGA-based implementation results for chaos-based secure image communications. **a** Two Altera DE2 FPGA development boards; **b** hardware platform; **c** case of mismatched parameter; **d** case of matched parameters. (©IEEE 2016, Reproduced with permission from [5])

ing principle of the hardware system is expressed as follows. In Fig. 5.12a, the picture is previously stored in the Altera DE2's RAM (random access memory), namely Picture RAM in Fig. 5.11. When the system is turned on, the picture is sent in two ways: one is delivered to the VGA display controller; it collects the RGB information of every pixel from the picture, and then transmits them to the monitor for display. The other way is sent to the RGB encryption module and the RGB elements of each pixel are encrypted by operating XOR with the output stream of 3D-DCS, respectively. This needs three different types of oscillator rings as external input sources. In addition, the control signal of states for three oscillator rings at the sender is also delivered through the public channel to synchronize those of 3D-DCS at the receiver.

In Fig. 5.12b, the received encrypted RGB objects are decrypted. Because the states of the 3D-DCS synchronize with those at the transmitter side, the RGB values of the picture could be recovered by executing XOR separately with the states at three dimensions of 3D-DCS. The RGB pixels are finally sent to the VGA controller and displayed on the monitor.

5.4.3 FPGA-Based Implementation Result for Image Encryption and Decryption

In the experiment, an image of 640×480 resolution was previously stored in the Altera DE2 board's RAM. The image encryption system based on 3D-DCS is implemented with this image. As shown in Fig. 5.11, only when the states of 3D-DCS between the transmitter and receiver are exactly matched can the original image be correctly decrypted; otherwise it cannot be recovered. The experimental FPGA-based implementation results are shown in Fig. 5.12c, d.

References

1. K. Hirano, T. Yamazaki, S. Morikatsu et al., Fast random bit generation with bandwidth-enhanced chaos in semiconductor lasers. Opt. Express **18**(6), 5512–5524 (2010)
2. T. Addabbo, A. Fort, L. Kocarev, S. Rocchi, V. Vignoli, Pseudo-chaotic lossy compressors for true random number generation. IEEE Trans. Circuits Syst. I **58**(8), 1897–1909 (2011)
3. M. Epstein, L. Hars, R. Krasinski, M. Rosner, H. Zheng, Design and Implementation of a True Random Number Generator Based on Digital Circuit Artifacts, vol. 2779, Lecture Notes in Computer Science (Springer, Berlin, 2003), pp. 152–165
4. B. Sunar, W. Martin, D. Stinson, A provably secure true random number generator with built-in tolerance to active attacks. IEEE Trans. Comput. **56**(1), 109–119 (2007)
5. Q. Wang, S. Yu, C. Li, J. Lü, X. Fang, C. Guyeux, J. Bahi, Theoretical design and FPGA-based implementation of higher-dimensional digital chaotic systems. IEEE Trans. Circuits Syst. I **63**(3), 401–412 (2016)
6. Q. Wang, S. Yu, C. Guyeux, J. Bahi, X. Fang, Theoretical design and circuit implementation of integer domain chaotic systems. Int. J. Bifurc. Chaos **24**(10) (2014). p. Art. no. 1450128
7. Q. Wang, S. Yu, C. Guyeux, J. Bahi, X. Fang, Study on a new chaotic bitwise dynamical system and its FPGA implementation. Chin. Phys. B **24**(6) (2015). Art. no. 60503
8. C.-S. Hsu, Global analysis by cell mapping. Int. J. Bifurc. Chaos **2**(4), 727–771 (1992)
9. A. Shreim, P. Grassberger, W. Nadler, B. Samuelsson, J.E.S. Socolar, M. Paczuski, Network analysis of the state space of discrete dynamical systems. Phys. Rev. Lett. **98**(19), 198701 (2007)
10. J. Banks, J. Brooks, G. Cairns, G. Davis, P. Stacey, On devaney's definition of chaos. Am. Math. Mon. **99**(4), 332–334 (1992)
11. C. Lin, G. Chen, Estimating the lyapunov exponents of discrete systems. Chaos **14**, 343–346 (2004)
12. S. Palnitkar, *Verilog HDL: A Guide to Digital Design and Synthesis*, 2nd edn. (Prentice Hall Professional, California, 2003)
13. R.J.J. Jenkins, Isaac. Fast Software Encryption **1039**, 41–49 (1996)
14. A. Rukhin, J. Soto, J. Nechvatal et al., A statistical test suite for random and pseudorandom number generators for cryptographic applications, NIST Special Publication 800-22rev1a (2010), http://csrc.nist.gov/groups/ST/toolkit/rng/documentation_software.html

Chapter 6
Investigating the Statistical Improvements of Various Chaotic Iterations-Based PRNGs

6.1 Various Algorithms for Pseudorandom Number Generation

After having investigated the theoretical study and the circuit design of our particular dynamical systems, we now investigate the software generation. We consider that the strategy is provided by a PRNG (pseudorandom number generator), leading to a collection of so-called "CIPRNGs" (chaotic iteration-based PRNGs) algorithms reviewed here.

6.1.1 Qualitative Relations Between Topological Properties and Statistical Tests

Let us first explain why we have reasonable ground to believe that chaos can improve the statistical properties of pseudorandom generator algorithms. We show in this section that chaotic properties as defined in the mathematical theory of chaos are related to some statistical tests that can be found in the NIST battery of tests. Later in this chapter we verify that, when mixing defective PRNGs with chaotic iterations, the new software generator presents better statistical properties.

There are indeed various relations between topological properties that describe an unpredictable behavior for a discrete dynamical system on the one hand, and statistical tests to check the randomness of a numerical sequence. These two mathematical disciplines follow a similar objective in the case of a recurrent sequence (to characterize an intrinsically complicated behavior) with two different but complementary approaches. It is true that the following illustrative links give only qualitative arguments, and proofs should be provided to make such arguments irrefutable. However, they give a first understanding of the reason why chaotic properties tend to improve the statistical quality of PRNGs, which is experimentally verified as shown at the end

Q. Wang et al., *Design of Digital Chaotic Systems Updated by Random Iterations*, SpringerBriefs in Nonlinear Circuits, https://doi.org/10.1007/978-3-319-73549-8_6

of this chapter. Let us now list some of these relations between topological properties defined in the mathematical theory of chaos and tests embedded in the NIST battery.

- **Regularity**. As stated earlier in this book, a chaotic dynamical system must have an element of regularity. Depending on the chosen definition of chaos, this element can be the existence of a dense orbit, the density of periodic points, and so on. The key idea is that a dynamical system with no periodicity is not as chaotic as a system having periodic orbits: in the first situation, we can predict something and gain knowledge about the behavior of the system; that is, it never enters into a loop. A similar importance for periodicity is emphasized in the two following NIST tests:

  - **Nonoverlapping Template Matching Test**. Detect generators that produce too many occurrences of a given nonperiodic (aperiodic) pattern.
  - **Discrete Fourier Transform (Spectral) Test**. Detect periodic features (i.e., repetitive patterns that are close to one another) in the tested sequence that would indicate a deviation from the assumption of randomness.

- **Transitivity**. This previously introduced topological property states that the dynamical system is intrinsically complicated: it cannot be simplified into two subsystems that do not interact, inasmuch as we can find in any neighborhood of any point another point whose orbit visits the whole phase space. As stated previously, this focus on the places visited by the orbits of the dynamical system takes various nonequivalent formulations in the mathematical theory of chaos, namely: transitivity, strong transitivity, total transitivity, topological mixing, and so on. A similar attention is brought on the states visited during a random walk in the two NIST tests below:

  - **Random Excursions Variant Test**. Detect deviations from the expected number of visits to various states in the random walk.
  - **Random Excursions Test**. Determine if the number of visits to a particular state within a cycle deviates from what one would expect for a random sequence.

- **Chaos according to Li and Yorke**. We recalled that two points of the phase space (x, y) define a Li-Yorke couple when

$$\lim sup_{n \to +\infty} d(f^{(n)}(x), f^{(n)}(y)) > 0$$

and

$$\lim inf_{n \to +\infty} d(f^{(n)}(x), f^{(n)}(y)) = 0,$$

meaning that their orbits always oscillate as the iterations pass. When a system is compact and contains an uncountable set of such points, it is claimed as chaotic according to Li-Yorke. A similar property is regarded in the following NIST test.

  - **Runs Test**. This is to determine whether the number of runs of ones and zeros of various lengths is as expected for a random sequence. In particular, this test determines whether the oscillation between such zeros and ones is too fast or too slow.

- **Topological Entropy**. The desire to formulate an equivalency of the thermodynamics entropy has emerged both in the topological and statistical fields. Once again, a similar objective has led to two different rewritings of an entropy-based disorder: the famous Shannon definition is approximated in the statistical approach, whereas topological entropy has been defined previously. This value measures the average exponential growth of the number of distinguishable orbit segments. In this sense, it measures the complexity of the topological dynamical system, whereas the Shannon approach comes to mind when defining the following test.

  – **Approximate Entropy Test**. Compare the frequency of the overlapping blocks of two consecutive/adjacent lengths (m and $m + 1$) against the expected result for a random sequence.

- **Nonlinearity, Complexity**. Finally, let us remark that nonlinearity and complexity are not only sought in general to obtain chaos, but they are also required for randomness, as illustrated by the two tests below.

  – **Binary Matrix Rank Test**. Check for linear dependence among fixed length substrings of the original sequence.
  – **Linear Complexity Test**. Determine whether the sequence is complex enough to be considered random.

We have recalled that chaotic iterations are, among other things, strongly transitive, topologically mixing, chaotic as defined by Li and Yorke, and that they have a topological entropy and an exponent of Lyapunov both equal to $ln(N)$, where N is the size of the iterated vector; see the first chapters of this book. Due to these topological properties, we are bound to believe that software based on chaotic iterations could probably be able to pass batteries for pseudorandomness such as the NIST one. The following subsections show that defective generators have their statistical properties improved by chaotic iterations.

6.1.2 CIPRNGs: Chaotic Iteration-Based PRNG Algorithms

This section focuses on the presentation of various realizations of pseudorandom number generators based on chaotic iterations. Each is a particular software implementation of the one-dimensional digital chaotic systems (ODDCS) system presented in the previous chapter.

6.1.2.1 CIPRNG, Version 1

Let $N \in \mathbb{N}^*$, $N \geqslant 2$, and $\mathscr{M}$ be a finite subset of $\mathbb{N}^*$. Consider two possibly defective generators called PRNG1 and PRNG2 we want to improve, the first one having its terms in $[\![1, N]\!]$ whereas the second returns integers in $\mathscr{M}$, which is always possible. The first version of a generator resulting in a posttreatment on these defective PRNGs using chaotic iterations is denoted by CIPRNG(PRNG1,PRNG2) version 1. This (inefficient) proof of concept is designed by the following process.

1. Some chaotic iterations are fulfilled, with the vectorial negation and PRNG1 as strategy, to generate a sequence $(x^n)_{n \in \mathbb{N}} \in (\mathbb{B}^\mathsf{N})^\mathbb{N}$ of Boolean vectors, where $\mathbb{B}$ is the set $\{0, 1\}$: the successive internal states of the iterated system.
2. Some of these vectors are randomly extracted with PRNG2 and their components constitute our pseudorandom bit flow. Algorithm 1 provides the way to produce one output.

Input: The internal state x (an array of N 1-bit words)
Output: An array of N 1-bit words
   1: **for** $i = 0, \ldots, PRNG1()$ **do**
   2:   $S \leftarrow PRNG2()$;
   3:   $x_S \leftarrow \overline{x_S}$;
   4: **end for**
   5: return x;

Algorithm 1: An arbitrary round of CIPRNG(PRNG1,PRNG2) version 1.

In other words, chaotic iterations are realized as follows. Initial state $x^0 \in \mathbb{B}^\mathsf{N}$ is a Boolean vector taken as a seed and strategy $(S^n)_{n \in \mathbb{N}} \in [\![1, \mathsf{N}]\!]^\mathbb{N}$ is a sequence produced by PRNG2. Lastly, the iterative function f is the vectorial Boolean negation. Thus, at each iteration, only the S^ith component of state x^n is updated, as follows.

$$
x_i^n = \begin{cases} x_i^{n-1} & \text{if } i \neq S^i, \\[2ex] \overline{x_i^{n-1}} & \text{if } i = S^i. \end{cases} \tag{6.1}
$$

Finally, some x^n are selected by a sequence m^n as the pseudorandom bit sequence of our generator, where $(m^n)_{n \in \mathbb{N}} \in \mathcal{M}^\mathbb{N}$ is obtained using PRNG2. That is, the generator returns the values: the components of x^{m^0}, followed by the components of $x^{m^0 + m^1}$, followed by the components of $x^{m^0 + m^1 + m^2}$, and so on.

Generators investigated in the first set of experiments are the logistic map, XOR-shift, and ISAAC (indirection, shift, accumulate, add, and count)[1], and the reputed NIST, DieHARD, and TestU01 test suites have been considered for statistical evaluation. Table 6.1 contains the statistical results obtained by the considered inputted generators, whereas Table 6.2 shows the results with the first version of our CIPRNGs: improvements are obvious.

---

[1] All the generators discussed in this chapter are defined in the appendix.

Table 6.1 Statistical results of well-known PRNGs

	BBS	Logistic	XORshift	ISAAC
NIST SP 800-22 (15 tests)	2	14	14	15
DieHARD (18 tests)	2	16	15	18
TestU01 (516 tests)	212	250	370	516

Table 6.2 Statistical results for the CIPRNG version 1

Test name	CIPRNG version 1			
	Logistic	XORshift	ISAAC	ISAAC
	+	+	+	+
	Logistic	XORshift	XORshift	ISAAC
NIST (15)	15	15	15	15
DieHARD (18)	18	18	18	18
TestU01 (516)	378	507	516	516

6.1.2.2 The CIPRNG Version 2

After the proof of concept of CIPRNG version 1, a second version of a generator based on chaotic iterations has been introduced in order to obtain outputs at the same speed as the inputted generators. The basic idea in this improvement is to prevent changing a bit twice between two outputs, and by doing so reducing the generation time. To do this, the meaning of sequence (m^n) must be changed: it now defines the number of bits to change between two outputs (instead of the number of chaotic iterations). This version is indeed the algorithm version of the IDCS presented in Chap. 3.

The output of the sequence PRNG2 is normally uniform in $[\![0, 2^{32} - 1]\!]$. However, we do not want the output of (m^n) to be uniform in $[\![0, N]\!]$, because in this case, the returns of our generator will not be uniform in $[\![0, 2^N - 1]\!]$, as illustrated in the following example. Suppose that $x^0 = (0, 0, 0)$. Then $m^0 \in [\![0, 3]\!]$: the number of bits we can change in x^0 is between 0 and 3.

- If $m^0 = 0$, then no bit must change between the first and the second output of our CIPRNG version 2. Thus we have only one possibility for x^1, namely $x^1 = (0, 0, 0)$.
- If $m^0 = 1$, then exactly one bit must change, which leads to three possible values for x^1, that is, $(1, 0, 0)$, $(0, 1, 0)$, and $(0, 0, 1)$.
- etc.

Each value in $[\![0, 2^3 - 1]\!]$ must be returned with the same frequency; then the values $(0, 0, 0)$, $(1, 0, 0)$, $(0, 1, 0)$, and $(0, 0, 1)$ must occur for x^1 with the same probability. Finally we see that, in this example, $m^0 = 1$ must be three times more probable than $m^0 = 0$. This leads to the general definition for the probability of $m = i$:

$P(m^n = i) = \frac{C_N^i}{2^N}$, with $C_n^k = \frac{n!}{k!(n-k)!}$. Thus $\forall n \in \mathbb{N}, m^n = g(PRNG2()^n)$,

where

Table 6.3 Statistical results for the CIPRNG version 2

Test name	CIPRNG Version 2			
	Logistic + Logistic	XORshift + XORshift	ISAAC + XORshift	ISAAC + ISAAC
NIST (15)	15	15	15	15
DieHARD (18)	18	18	18	18
TestU01 (516)	516	516	516	516

$$g(y) = k \Leftrightarrow \sum_{i=0}^{k-1} C_N^i \leqslant y < \sum_{i=0}^{k} C_N^i.$$

We have adapted the outputs of PRNG2 to obtain a sequence corresponding to the number of changes between two outputs of the CIPRNG(PRNG1,PRNG2) version 2. We must also adapt the outputs of the strategy PRNG1: the strategy indicates the coordinate change, and we do not want to change a given coordinate twice between two outputs of the CIPRNG. More precisely, the m^0 first terms of the strategy must be different, because we want to obtain m^0 changes between the initial state and the first output. Then, the terms in position m^0+1, ..., m^1 must be all different as well for the same reason, and thus one. However, PRNG1 does not necessarily provide such a particular sequence. This is why we must operate a decimation on it, as follows.

Let $(d^1, d^2, \ldots, d^N) \in \{0, 1\}^N$ be a mark sequence, counting the number of occurrences of each integer between two outputs. It is such that whenever $\sum_{i=1}^{N} d^i = m^k$, then $\forall i, d_i = 0$: after each output of the CIPRNG, this counting sequence is reset. This mark sequence will control the PRNG1 sequence b as follows. Let b^j be the numbers produced by PRNG1:

- If $d^{b^j} \neq 1$, then $S^k = b^j$, $d^{b^j} = 1$, and $k = k + 1$.
- If $d^{b^j} = 1$, then b^j is discarded (it has already occurred in this output).

The basic design procedure of this optimized generator is summed up in Algorithm 2, whereas good obtained results are provided in Table 6.3.

6.1.2.3 XOR CIPRNG

Instead of updating only one cell at each iteration as in the previous versions of our CIPRNGs, we can try to choose a subset of components and to update them together. Such an attempt leads to a kind of merger of the two random sequences. When the updating function is the vectorial negation, this algorithm can be rewritten as follows.

$$\begin{cases} x^0 \in [\![0, 2^N - 1]\!], \, S \in [\![0, 2^N - 1]\!]^{\mathbb{N}} \\ \forall n \in \mathbb{N}^*, \, x^n = x^{n-1} \oplus S^n, \end{cases} \tag{6.2}$$

Input: the internal state x (N bits)
Output: a state r (N bits)

   1: **for** $i = 0, \ldots, N$ **do**
   2:    $d_i \leftarrow 0$;
   3: **end for**
   4: $i \leftarrow 0$;
   5: $m \leftarrow g(PRNG2())$;
   6: **while** $i < m$ **do**
   7:    $S \leftarrow PRNG1()$;
   8:    **if** $d_S = 0$ **then**
   9:      $x_S \leftarrow \overline{x_S}$;
 10:     $d_S \leftarrow 1$;
 11:     $i \leftarrow i + 1$;
 12:    **end if**
 13: **end while**
 14: **return** x;

Algorithm 2: An arbitrary round of the CIPRNG(PRNG1,PRNG2) version 2.

and this rewriting can be understood as follows. The nth term S^n of the sequence S, which is an integer of N binary digits, whose list of digits in binary decomposition is the list of cells to update in the state x^n of the system (represented as an integer having N bits too). More precisely, the kth component of this state (a binary digit) changes if and only if the kth digit in the binary decomposition of S^n is 1. This generator has been called XOR CIPRNG and its chaos property has already been stated earlier in this book. It uses a very classical pseudorandom generation approach; the unique contribution is its relation with chaotic iterations: the single basic component presented in Eq. 6.2) is of ordinary use as a good elementary brick in various PRNGs. It corresponds to the discrete dynamical system in chaotic iterations.

6.2 On the Statistical Improvements of CIPRNG Posttreatments

6.2.1 First Investigations

The chaotic iterations-based generators proposed in this book are experimented here on various inputted pseudorandom generators, most of them being more or less defective. The objective is to show that, using chaos, we can improve the statistical profile of these defective generators. These well-known PRNGs have been considered for experiments:

- LCG, MRG for linear congruential PRNGs
- AWC, SWB, SWC, and GFSR for lagged ones
- INV from type ICG (inversive congruential generators)
- 2LCG, 3LCG, and 2MRG to study the effects on mixed PRNGs

Table 6.4 NIST and DieHARD tests suite passing rates for PRNGs without CIPRNG method

	Linear		Lagged				icg	Mixed		
	lcg	mrg	awc	swb	swc	gfsr	inv	2lcg	3lcg	2mrg
NIST	$\frac{11}{15}$	$\frac{14}{15}$	$\frac{15}{15}$	$\frac{15}{15}$	$\frac{14}{15}$	$\frac{14}{15}$	$\frac{14}{15}$	$\frac{14}{15}$	$\frac{14}{15}$	$\frac{14}{15}$
DieHARD	$\frac{16}{18}$	$\frac{16}{18}$	$\frac{15}{18}$	$\frac{16}{18}$	$\frac{18}{18}$	$\frac{16}{18}$	$\frac{16}{18}$	$\frac{16}{18}$	$\frac{16}{18}$	$\frac{16}{18}$

Table 6.5 NIST and DieHARD tests suite passing rates for PRNGs with CIPRNG method

Types	Linear		Lagged				icg	Mixed		
	lcg	mrg	awc	swb	swc	gfsr	inv	2lcg	3lcg	2mrg
Version 1										
NIST	$\frac{15}{15}$ *	$\frac{15}{15}$ *	$\frac{15}{15}$	$\frac{15}{15}$	$\frac{15}{15}$ *	$\frac{15}{15}$ *	$\frac{15}{15}$ *	$\frac{15}{15}$ *	$\frac{15}{15}$ *	$\frac{15}{15}$
DieHARD	$\frac{18}{18}$ *	$\frac{18}{18}$ *	$\frac{18}{18}$ *	$\frac{18}{18}$ *	$\frac{18}{18}$	$\frac{18}{18}$ *	$\frac{18}{18}$ *	$\frac{18}{18}$ *	$\frac{18}{18}$	$\frac{18}{18}$ *
Version 2										
NIST	$\frac{15}{15}$ *	$\frac{15}{15}$ *	$\frac{15}{15}$	$\frac{15}{15}$	$\frac{15}{15}$ *	$\frac{15}{15}$ *	$\frac{15}{15}$ *	$\frac{15}{15}$ *	$\frac{15}{15}$ *	$\frac{15}{15}$
DieHARD	$\frac{18}{18}$ *	$\frac{18}{18}$ *	$\frac{18}{18}$ *	$\frac{18}{18}$ *	$\frac{18}{18}$	$\frac{18}{18}$ *	$\frac{18}{18}$ *	$\frac{18}{18}$ *	$\frac{18}{18}$ *	$\frac{18}{18}$ *
XOR ciprng										
NIST	$\frac{14}{15}$ *	$\frac{15}{15}$ *	$\frac{15}{15}$	$\frac{15}{15}$	$\frac{14}{15}$	$\frac{15}{15}$ *	$\frac{14}{15}$	$\frac{15}{15}$ *	$\frac{15}{15}$ *	$\frac{15}{15}$
DieHARD	$\frac{16}{18}$	$\frac{16}{18}$	$\frac{17}{18}$ *	$\frac{18}{18}$ *	$\frac{18}{18}$	$\frac{18}{18}$ *	$\frac{16}{18}$	$\frac{16}{18}$	$\frac{16}{18}$	$\frac{16}{18}$

These defective generators are detailed, for instance, in the TestU01 manual. We have performed some statistical tests on these generators, showing that they reveal several issues, as summarized in Table 6.4. The tests studied here are the NIST suite and DieHARD battery of tests. Then we compared these results with the ones obtained after a chaotic iterations posttreatment.

We performed statistical analyses on each of the aforementioned CIPRNGs. The results are reproduced in Table 6.5. An asterisk "*" means that the considered passing rate has been improved. We can observe that, except for the XOR CIPRNG, all of the CIPRNGs have passed the 15 tests of the NIST battery and the 18 tests of the DieHARD one. Moreover, considering these scores, we can deduce that both the single Version 1 CIPRNG and the single Version 2 CIPRNG are relatively steadier than the single XOR CIPRNG approach when applying them to different PRNGs. However, the XOR CIPRNG is obviously the fastest approach to generate a CI random sequence, and it still improves the statistical properties relative to each generator taken alone, although the test values are not as good as desired.

To achieve a realization of the XOR CIPRNG that can pass all the tests embedded into the NIST battery, we now investigate the "Multiple XOR CIPRNG" variations of the XOR posttreatment in the following sections.

Table 6.6 Functional power m making it possible to pass the whole NIST battery

Inputted *PRNG*	lcg	mrg	swc	gfsr	inv	2lcg	3lcg	2mrg
Threshold value m	19	7	2	1	11	9	3	4

6.2.2 Variations on the XOR CIPRNG

We regard the possibility of using various successive terms of a given deficient generator S in order to improve its statistics. Such a desire leads to the definition of the multiple XOR CIPRNG detailed below:

$$\begin{cases} x^0 \in [\![0, 2^N - 1]\!], \, S \in [\![0, 2^N - 1]\!]^{\mathbb{N}} \\ \forall n \in \mathbb{N}^*, \, x^n = x^{n-1} \oplus S^{nm} \oplus S^{nm+1} \ldots \oplus S^{nm+m-1}, \end{cases} \tag{6.3}$$

where S stands for the inputted PRNG. We show in Table 6.6 that a threshold value m (called the functional power) can always be found such that the multiple XOR CIPRNG becomes able to pass the whole NIST battery. The existence of this threshold illustrates to a certain extent the progressive appearance of the effects of chaos.

The results presented in this section reinforce our confidence in the capability for chaos to act as posttreatment on defective pseudorandom number generators in order to improve their statistics. However, we still regret the following flaws for all the currently proposed CIPRNGs.

1. Up to now, speed performances are bad, inasmuch as in (single) CIPRNGs versions 1 and 2 we must call various times the inputted generators between two outputs. Similarly the XOR CIPRNG can satisfactorily improve defective generators only by grouping (XORing) a potentially large number of successive terms produced by the input (this is the multiple XOR CIPRNG).
2. As presented here, XOR and multiple XOR can only handle one inputted generator. However, an interesting strategy when designing new generators using formerly released ones is to take the best of each input: speed of the first inputted PRNG and security of the second one, for instance.
3. CIPRNGs versions 1 and 2 and multiple XOR CIPRNGs have better statistical performances than XOR CIPRNG, because they use various successive terms of the inputs to produce one output: chaos has time to express itself and high correlations between two successive inputs of the deflated PRNGs are broken by doing so.

We thus introduce two new methods to take the best of each version. They are described hereafter.

6.2.3 *"LUT" CIPRNG (XORshift, XORshift) Version 3*

The LUT (lookup-table) CIPRNG version 3 is an improved mixed version of both the CIPRNG version 2 and the XOR CIPRNG. The key ideas are:

1. To use a lookup table for faster production of strategies than in CIPRNG version 2 These strategies satisfy the same property as the ones provided by the decimation process, reducing by doing so the correlations of successive terms in the inputted PRNG.
2. To operate as in XOR CIPRNG, by computing $x^{n+1} = x^n \oplus S^n$ directly (general chaotic iterations of the vectorial negation instead of unary chaotic iterations)

This generator is not explained in detail in this chapter; only statistical test results are presented here.

6.2.4 *The Version 4 Category of CIPRNGs*

The CIPRNG version 4 is an improvement of the multiple XOR CIPRNG, in which we use m PRNGs instead of m successive terms of one PRNG, or, more precisely, subsets of these m PRNGs. By doing so, the problem of speed can be resolved by computing them in parallel, whereas the two other issues will no longer be problems.

In the XOR CIPRNG $x^{n+1} = x^n \oplus S^n$, the kth component of its state (a binary digit) changes if and only if the kth digit in the binary decomposition of the nth term S^n of the inputted generator is 1. In this algorithm, instead of updating only one cell at each iteration as the first versions of our CIPRNGs, a subset of components is chosen and updated. We have already shown that, taken alone, this XOR CIPRNG does not greatly improve the possibly defective inputted generator S. A first solution has been proposed in the multiple XOR CIPRNG by XORing various successive terms of S before XORing the result with the last state of the system. We have shown that this method is truly able to improve the inputted generator. However, its principal flaw is that, for the majority of generators, all the terms $S^{mn}, S^{mn+1}, …, S^{mn+m-1}$ must be computed step by step, and m can be large for very defective PRNGs. A second but less critical flaw is that the XOR CIPRNG only receives one inputted generator. However, as stated before, some situations exist where we want to benefit from various inputted generators: the security of the first PRNG, speed of the second one, and so on.

It is possible to add more complexity and speed in the multiple XOR CIPRNG, by considering a set of M inputted generators, picking a subset of them randomly at each iteration, and XORing their XORed values with the internal state of the system. This algorithm can be written as in Algorithm 3.

$S(1), S(2), …, S(M)$ are the M inputted PRNGs, whereas $T^n \in [\![0, 2^M - 1]\!]$ gives which sequences must be considered at the current iteration, as follows.

Let $(t_1^n, t_2^n, …, t_M^n) \in \{0, 1\}^M$ be the binary representation of the M-bit number T^n. Then the sequence $S^n(1), S^n(2), …, S^n(M)$ is decimated with the h function as

Input: the internal state x (N bits)
Output: a state r of N bits
 1: **for** $i = 1, \ldots, M$ **do**
 2:    $S(i) = PRNG2\_i()$;
 3: **end for**
 4: $T = PRNG1()$;
 5: $r = x \oplus h(T, S(1), S(2), \ldots S(M))$,
 6: **return** r;

Algorithm 3: An arbitrary round of the version 4 CI generator.

follows. if $t_i^n = 0$, then $S^n(i)$ is discarded, else $S^n(i)$ is kept for *bitwise exclusive or* computing. In brief, the produced output sequence x^n, based on chaotic iterations, is updated by a *bitwise exclusive OR* of an irregular decimation of $S(1)$, $S(2)$, ..., $S(M)$, according to the bits of T^n.

6.2.5   Randomness Quality of CIPRNGs

In this section, we investigate more deeply the statistical improvement obtained by realizing the various CIPRNG-based posttreatments introduced previously. Table 6.7 compares all the versions of CIPRNG (XORshift, XORshift) against the NIST and DieHARD batteries. We can see that the XORshift alone fails both batteries, whereas the generator based on discrete chaotic iterations (CIPRNG versions 1-4) can improve them.

Generators investigated in this second set of experiments are now, respectively, the BBS (with very bad security parameters: m of 32 bits and outputs of 4 bits), a logistic map, XORshift, and ISAAC (details of these generators are provided in the appendix), and the NIST, DieHARD, and TestU01 test suites have been considered for statistical evaluation. Let us recall that Table 6.1 contains the statistical results obtained by the considered inputted generators. In Table 6.8 are shown the results obtained by version 3 of our CIPRNGs. These results confirm that the CIPRNGs version 3 are all able to pass these tests, except when using the very deflated BBS generator. This issue is solved with version 4, as shown in Table 6.9. This last version of the CIPRNG family

Table 6.7 NIST and DieHARD results for XORshift alone and CIPRNG (XORshift, XORshift) versions 1-4

Version	XORshift	CIPRNG			
	Alone	1	2	3	4
NIST (15 tests)	14	15	15	15	15
DieHARD (18 tests)	15	18	18	18	18

Table 6.8   Statistical results for the LUT CIPRNG version 3

Test name	LUT CIPRNG Version 3			
	Logistic	XORshift	ISAAC	BBS
	+	+	+	+
	Logistic	XORshift	XORshift	XORshift
NIST (15)	15	15	15	8
DieHARD (18)	18	18	18	8
TestU01 (516)	516	516	516	356

Table 6.9   Statistical results for the CIPRNG version 4

Test name	CIPRNG version 4			
	Logistic	XORshift	ISAAC	BBS
	+	+	+	+
	Logistic	XORshift	XORshift	XORshift
NIST (15)	15	15	15	15
DieHARD (18)	18	18	18	18
TestU01 (516)	516	516	516	516

thus offers a great compromise between statistical performance and efficiency. It can be considered very suitable for both software and hardware implementations.

6.3   Practical Security Evaluation

Given a key size, it is possible to measure in practice the minimum duration needed for an attacker to break a cryptographically secure PRNG, if we know the power of his or her machines. Such a concrete security evaluation is related to the (T, ε)−security notion, which is recalled and evaluated in what follows, for the sake of completeness.

Let us first recall that,

Definition 6.1   Let $\mathscr{D} : \mathbb{B}^M \longrightarrow \mathbb{B}$ be a probabilistic algorithm that runs in time T. Let $\varepsilon > 0$. $\mathscr{D}$ is called a (T, ε)−distinguishing attack on pseudorandom generator G if

$$\left| Pr[\mathscr{D}(G(k)) = 1 \mid k \in_R \{0, 1\}^{\ell}] - Pr[\mathscr{D}(s) = 1 \mid s \in_R \mathbb{B}^M] \right| \geqslant \varepsilon,$$

where the probability is taken over the internal coin flips of $\mathscr{D}$, and the notation "$\in_R$" indicates the process of selecting an element at random and uniformly over the corresponding set.

Let us recall that the running time of a probabilistic algorithm is defined to be the maximum of the expected number of steps needed to produce an output, maximized over all inputs; the expected number is averaged over all coin flips made by the algorithm. We are now able to define the notion of cryptographically secure PRNGs:

Definition 6.2 A pseudorandom generator is (T, ε)−secure if there exists no (T, ε)−distinguishing attack on this pseudorandom generator.

Suppose now that the XOR CIPRNG with the cryptographically secure BBS as input will work during $M = 100$ time units, and that during this period, an attacker can realize 10^{12} clock cycles. We thus wonder whether, during the PRNG's lifetime, the attacker can distinguish this sequence from a truly random one, with a probability greater than $\varepsilon = 0.2$. We consider that the modulus of BBS N has 900 bits, that is, contrary to the simulations section, we use here the BBS generator with relevant security parameters.

Predicting the next generated bit knowing all the previously released ones by the XOR CIPRNG is obviously equivalent to predicting the next bit in the BBS generator, which is cryptographically secure. More precisely, it is (T, ε)−secure: no (T, ε)−distinguishing attack can be successfully realized on this PRNG, if

$$T \leqslant \frac{L(N)}{6N(log_2(N))\varepsilon^{-2}M^2} - 2^7 N\varepsilon^{-2}M^2 log_2(8N\varepsilon^{-1}M), \qquad (6.4)$$

where M is the length of the output ($M = 100$ in our example), and $L(N)$ is equal to

$$2.8 \times 10^{-3} exp\left(1.9229 \times (N \ln 2)^{\frac{1}{3}} \times (ln(N \ln 2))^{\frac{2}{3}}\right)$$

is the number of clock cycles to factor an N−bit integer.

A direct numerical application shows that this attacker cannot achieve a $(10^{12}, 0.2)$ distinguishing attack in that context.

Chapter 7
Conclusions

To solve the degradation of chaotic dynamical properties caused by limitation of finite-precision presentation and quantization, this book developed digital chaotic systems updated by random iterations from low- to high-dimensional settings, utilizing the chaos generation strategy controlled by random sequences. After having recalled the bases of iterative systems, chaotic iterations, and the mathematical theory of chaos, the way to study iterative systems in this mathematical framework has then been explained, and previously obtained conditions for chaos have then been listed. The way to compute them without loss of chaos has then been explained: the problem of finite state machines and the way to evaluate chaos of a given program have been solved. Finally, the effects of mixing defective PRNGs (pseudorandom number generators) using chaotic iterations has been largely regarded, by recalling our previously obtained improvements of their statistics, for various ways to operate the chaotic iterations-based posttreatment we called CIPRNG (chaotic iteration-based PRNG).

This research allows the construction of both machines and computer programs having unpredictable behaviors. Existing algorithms can also be studied using the mathematical topology framework, in order to compare them or to discover new threats. Finally, new topological properties can be added to existing tools with the proposed posttreatment without loss of security. Motivations to use chaotic programs are manifold: to be in good condition when designing new algorithms, to create new kinds of attacks such as chaotic viruses, to simulate chaotic processes numerically, to create neural networks proven as chaotic, to reinforce the security of schemes already proven as cryptographically secure, or to struggle with artificial intelligence that is used, for instance, in steganalizers (we have shown that neural networks fail in learning some chaotic iteration behaviors), to name a few.

Further investigations encompass the study of the choice of topology when comparing the quality of two different programs, or to have an absolute scale to evaluate an algorithm. Situations where the inputted generator is a true random number

Q. Wang et al., *Design of Digital Chaotic Systems Updated by Random Iterations*, SpringerBriefs in Nonlinear Circuits, https://doi.org/10.1007/978-3-319-73549-8_7

generator (TRNG) must also be deepened by more thoroughly investigating the analog/numerical mixture, in which our CIPRNG (chaotic iteration-based PRNG) receives the output of a chaotic optoelectronic laser. Then, the correlations between some statistical tests and some topological properties must be systematically investigated.

Open problems in the field of digital chaotic systems encompass the understanding of the links between randomness and chaos on the one hand, and security and chaos. Quantitative measurements of chaotic behaviors, by the mean of measure theory, should be investigated too, by evaluating the ergodicity and the weak mixing of such discrete dynamical systems, depending on the embedded iterative function. Kolmogorov complexity has to be regarded as well during these quantitative measurements of chaos. Finally, the quality of the chaos generated by a computer system should be compared to physical chaos and to white noise, and tools allowing such comparisons have to be designed.

Appendix A
Some Well-Known Generators

We now introduce various well-known pseudorandom number generators. They have been used previously in this book, to evaluate the quality of a posttreatment based on chaotic iterations.

A.1 Blum Blum Shub

The Blum Blum Shub generator (usually denoted by BBS) takes the form:

$$x^0 \in [\![1, m-1]\!]$$

$$x^{n+1} = (x^n \times x^n) \bmod m, \quad y^{n+1} = x^{n+1} \bmod (log(m)),$$

where m is the product of two prime numbers p and q, such that:

- p and q are congruent to 3 modulus 4
- $gcd(\phi(p-1), \phi(q-1))$ should be small[1]

y^n is the returned sequence, whereas log refers to the logarithm to base 2.

This generator is known to be secure for sufficiently large p and q. However, in this book, we do not focus on security, but on statistical improvement of defective generators: we want to show that deficient PRNGs (pseudorandom number generators) can be improved using the chaotic iterations posttreatment. A way to find such defective generators is to use good ones like this BBS but in a wrong context (small prime numbers, in this situation).

---

[1] Euler's totient $\phi(n)$ is an arithmetic function that counts the number of positive integers less than or equal to n that are relatively prime to n.

© The Author(s), under exclusive licence to Springer International Publishing AG, part of Springer Nature 2018
Q. Wang et al., *Design of Digital Chaotic Systems Updated by Random Iterations*, SpringerBriefs in Nonlinear Circuits, https://doi.org/10.1007/978-3-319-73549-8

A.2 Logistic Map

The logistic map is given by:

$$x^{n+1} = \mu\, x^n (1 - x^n), \text{ with } x^0 \in (0, 1),\ \mu \in [0, 4],$$

where x is a real number. The logistic map was originally introduced as a demographic model by Pierre François Verhulst in 1838. In 1947, Ulam and Von Neumann studied it as a PRNG. This essentially requires mapping the states of the system $(x^n)_{n \in \mathbb{N}}$ to $\{0, 1\}^{\mathbb{N}}$. A simple way for turning x^n to a discrete bit symbol r is by using a threshold function as shown in Algorithm 4. A second usual way to obtain an integer sequence from a real system is to chop off the leading bits after moving the decimal point of each x to the right, as obtained in Algorithm 5.

Input: the internal state x (a decimal number)
Output: r (a 1-bit word)
1: $x \leftarrow 4x(1 - x)$
2: **if** $x < 0$ **then**
3:     $r \leftarrow 0$;
4: **else**
5:     $r \leftarrow 1$;
6: **end if**
7: return r;

Algorithm 4: An arbitrary round of logistic map 1

The logistic map is a famous example of Devaney's chaotic dynamical system for $\mu \in (3.99996, 4]$. However, it is statistically biased and its implementation on machines with finite precision raises many problems. In this book, we have used it with the method of Algorithm 4 and a threshold equal to 0.5.

Input: the internal state x (a decimal number)
Output: r (an integer)
1: $x \leftarrow 4x(1 - x)$
2: $r \leftarrow \lfloor 10000000x \rfloor$
3: return r;

Algorithm 5: An arbitrary round of logistic map 2

A.3 Linear Congruential Generator

The linear congruential generator (LCG) is defined by the recurrence:

$$x^0 \in [0, m-1], \quad x^n = (ax^{n-1} + c) \bmod m \tag{A.1}$$

where a, c, and x^0 are positive integers less than m, called, respectively, the multiplier, increment, and seed of the generator [1]. LCG is one of the oldest and best-known generators. It will have a full period for all seed values if and only if:

1. c and m are relatively prime.
2. $a - 1$ is divisible by all prime factors of m.
3. $a - 1$ is a multiple of 4 when m is a multiple of 4.

In this book, 2LCGs and 3LCGs refer to two (resp., three) combinations [1] of such LCGs, as follows.

- The first LCG s_1 has parameters (m_1, a_1, c_1) and the second one s_2 has parameters (m_2, a_2, c_2),
- The combination is $x^n = (s_1^n - s_2^n) \bmod (m_1 - 1)$, where s_1^n and s_2^n are the states of the two LCGs components at step n.

In other words:

$$\begin{cases} s_1^n = (a_1 \times s_1^{n-1} + c_1) \bmod m_1 \\ s_2^n = (a_2 \times s_1^{n-1} + c_2) \bmod m_2 \\ x^n = (s_1^n - s_2^n) \bmod (m_1 - 1). \end{cases}$$

These formulas can be easily adapted for the combination of three linear congruential generators. The inputted LCGs must satisfy the requirement mentioned above, and one must also have $m1 > m2$.

A.4 Multiple Recursive Generators

The multiple recursive generators (MRGs) are based on higher-order recursion k [1]:

$$x^0 \in [0, m-1], \quad x^n = (a^1 x^{n-1} + \cdots + a^k x^{n-k}) \bmod m, \tag{A.2}$$

where $a^1, \ldots, a^k \in [0, m-1]$. Combination of two MRGs (referred to as 2MRGs) has also been used in this book; they are defined like the multiple LCGs:

$$\begin{cases} s_1^n = (a_1^1 s_1^{n-1} + \cdots + a_1^k s_1^{n-k}) \bmod m_1, \\ s_2^n = (a_2^1 s_2^{n-1} + \cdots + a_2^k s_2^{n-k}) \bmod m_2, \\ x^n = (s_1^n - s_2^n) \bmod (m_1). \end{cases}$$

The combination method is thus obtained by subtracting the states modulo m_1. For reasons not debated in this book, usual implementations of this 2MRG that present correct statistics suppose that $k = 3$, $a_1^1 = 0$, $a_1^2 > 0$, $a_1^3 < 0$, $a_2^1 > 0$, $a_2^2 = 0$, $a_2^3 < 0$, and finally $a_1^j \times (m_1 \bmod a_1^j) < m_1$ whereas $a_2^j \times (m_2 \bmod a_2^j) < m_2$. These requirements have been followed in our experiments.

A.5 UCARRY

The UCARRY acronym refers to generators based on linear recurrences with carry. This includes the *add-with-carry* (AWC), *subtract-with-borrow* (SWB), and *shift-with-carry* (SWC) generators.

The add-with-carry generator, proposed by Marsaglia and Zaman, is based on the following linear recurrence with carry. Given a modulus m and two different positive integers r and s, and for integer initial values $c^0 \in \{0, 1\}$ and $x^0, \ldots, x^k \in [\![0, m-1]\!]$, where $k = max(r, s)$, compute for $n > k$:

$$
\begin{aligned}
x^n &= (x^{n-r} + x^{n-s} + c^{n-1}) \bmod m, \\
c^n &= (x^{n-r} + x^{n-s} + c^{n-1})/m,
\end{aligned}
\tag{A.3}
$$

and return at each iteration the output x^n/m, that is, their quotient.

The SWB generator, for its part, uses the same inputs and has the recurrence:

$$
\begin{aligned}
x^n &= (x^{n-r} - x^{n-s} - c^{n-1}) \bmod m, \\
c^n &= \begin{cases} 1 & \text{if } (x^{i-r} - x^{i-s} - c^{i-1}) < 0 \\ 0 & \text{else,} \end{cases}
\end{aligned}
\tag{A.4}
$$

and the output is x^i/m another time.

Finally, the shift-with-carry SWC generator designed by R. Couture is based on the following recurrence.

$$
\begin{aligned}
x^n &= (a^1 x^{n-1} \oplus \cdots \oplus a^r x^{n-r} \oplus c^{n-1}) \bmod 2^w, \\
c^n &= (a^1 x^{n-1} \oplus \cdots \oplus a^r x^{n-r} \oplus c^{n-1}) / 2^w.
\end{aligned}
\tag{A.5}
$$

with output equal to $x^n/2^w$. The initial values are $(x^0, \ldots, x^{r-1})$ and c is the initial carry with restrictions: $0 < r$, and $w \le 32$.

A.6 Generalized Feedback Shift Register

By GFSR we referred to a particular generalized feedback shift register generator based on the recurrence:

$$
x^n = x^{n-r} \oplus x^{n-k}.
\tag{A.6}
$$

Each x^n is a 32-bit vector and k and r are positive integers such that $r < k$. The output at step n is $u_n = \tilde{x}_n/2^l$, where $\tilde{x}_n$ is the integer formed by the first l bits of x_n, and $l \le 32$. $x_0, \ldots, x_{k-1}$ must be provided as k initial bit vectors. Proper initialization techniques for this generator have been discussed in the literature; they have been respected during our implementations.

A.7 Nonlinear Inversive Generator

Finally, INV stands for the nonlinear inversive generator, as defined in [1], which is:

$$x^n = \begin{cases} (a^1 + a^2/z^{n-1}) \bmod m & \text{if } z^{n-1} \neq 0 \\ a^1 & \text{if } z^{n-1} = 0. \end{cases} \tag{A.7}$$

The generator computes z via the modified Euclid algorithm (see [1]). If m is prime and if $p(x) = x^2 - a^1 x - a^2$ is a primitive polynomial modulo m, then the generator has maximal period m with restrictions: $0 \leq z^0 < m, 0 < a^1 < m$ and $0 < a^2 < m$. Furthermore, m must be a prime number, preferably large.

A.8 XORshift

XORshift is a category of very fast PRNGs designed by George Marsaglia. It repeatedly uses the transform of *exclusive OR* (XOR) on a number with a bit-shifted version of it.

The state of an XORshift generator is a vector of bits. At each step, the next state is obtained by applying the following operations three times to w-bit blocks in the current state, where $w = 32$ or 64: replace the w-bit block by a bitwise XOR of the original block with a shifted copy of itself by, respectively, a, b, and c positions, where $-w < a, b, c < w$. The direction of the circular shift is either the left or the right, depending on the signs of a, b, and c.

For instance, Algorithm 6 is the 32-bit XORshift with $(a, b, c) = (-13, 17, -5)$, which has a period of $2^{32} - 1 \approx 4.29 \times 10^9$.

Input: the internal state z (a 32-bits word)
Output: y (a 32-bits word)
1: $z \leftarrow z \oplus (z \ll 13)$;
2: $z \leftarrow z \oplus (z \gg 17)$;
3: $z \leftarrow z \oplus (z \ll 5)$;
4: $y \leftarrow z$;
5: return y;

Algorithm 6: An arbitrary round of XORshift algorithm

In this book, we have always supposed that the directions of the circular shifts are: left/right/left, which was not required in the original paper of Marsaglia. Other improved versions of this XORshift exist in the literature; in our research we have chosen this historical one for its speed and statistical flaws.

A.9 ISAAC

ISAAC is an array-based PRNG and a stream cipher designed by Robert Jenkins (1996) to be cryptographically secure. The name is an acronym for indirection, shift, accumulate, add, and count. The ISAAC algorithm has similarities with RC4. It uses an array of 256 32-bit integers as the internal state, writes the results to another 256-integer array, from which they are read one at a time until empty, at which point they are recomputed. Because it only takes about 19 32-bit operations for each 32-bit output word, it is extremely fast on 32-bit computers.

We give the key-stream procedure of ISAAC in Algorithm 7. The internal state is x, the output array is r, and the inputs 32-bit words a, b, and c are those computed in the previous round. Normally a, b, c, and the array x are initialized with some random sequences. The value $f(a, i)$ in Algorithm 7 is a 32-bit word, defined for all a and $i \in \{0, \ldots, 255\}$ as

$$f(a, i) = \begin{cases} a \ll 13 & \text{if } i \bmod 4 \equiv 0, \\ a \gg 6 & \text{if } i \bmod 4 \equiv 1, \\ a \ll 2 & \text{if } i \bmod 4 \equiv 2, \\ a \gg 16 & \text{if } i \bmod 4 \equiv 3. \end{cases} \tag{A.8}$$

Input: a, b, c, and the internal state x, they are 32-bit words
Output: an array r of 256 32-bit words
1: $c \leftarrow c + 1$;
2: $b \leftarrow b + c$;
3: **while** $i = 0, \ldots, 255$ **do**
4:      $s \leftarrow x_i$;
5:      $a \leftarrow f(a, i) + x_{(i+128) \bmod 256}$;
6:      $x_i \leftarrow a + b + x_{(x \gg 2) \bmod 256}$;
7:      $r_i \leftarrow s + x_{(x_i \gg 10) \bmod 256}$;
8:      $b \leftarrow r_i$;
9: **end while**
10: return r;

Algorithm 7: An arbitrary round of ISAAC algorithm

Reference

1. R. Simard, U.D. Montréal, Testu01: A software library in ANSI C for empirical testing of random number generators. ACM Trans. Math. Softw. **33**(4), 22 (2007)